Maiden Voyages

Also by Siân Evans

Queen Bees
Life Below Stairs
The Manor Reborn
Mrs Ronnie

UJ12
11|20

Maiden Voyages

*Women and the Golden Age of
Transatlantic Travel*

Siân Evans

First published in Great Britain in 2020 by Two Roads
An Imprint of John Murray Press
An Hachette UK company

I

A CIP catalogue record for this title is available from the British Library

Hardback ISBN 978 1 473 69902 1
Trade Paperback ISBN 978 1 473 69903 8
eBook ISBN 978 1 473 69905 2

Typeset in Sabon MT Std by Palimpsest Book Production Limited,
Falkirk, Stirlingshire

Printed and bound in Great Britain by Clays Ltd, Elcograf S.p.A.

John Murray Press policy is to use papers that are natural, renewable
and recyclable products and made from wood grown in sustainable
forests. The logging and manufacturing processes are expected to
conform to the environmental regulations of the country of origin.

Two Roads
Carmelite House
50 Victoria Embankment
London EC4Y 0DZ

www.tworoadsbooks.com

For my sister Sarah, best friend and fellow traveller.

Cunard Chief Officer Stephen Gronow of the *Aquitania* was
the author's great-great uncle. Researching his story, and that of his
shipmates, male and female, led to this book.

Contents

Dramatis Personae

Josephine Baker (1906–75)
African-American dancer and singer of consummate ability.
Born in St Louis, Missouri, she was talent-spotted in New
York and in 1925 she sailed to France to perform with other
black artistes in *La Revue Nègre*. Josephine's talent and
personality shone in Paris, her adopted city.

Tallulah Bankhead (1902–68)
American-born actress of original and startling talent, whose
stage career on both sides of the Atlantic gained considerable
acclaim, but whose everyday behaviour was outrageous. She
benefited from the symbiotic nature of British and American
showbusiness and society in the 1920s.

Victoria Drummond, MBE (1894–1978)
Named after her godmother, Queen Victoria, she was a
pioneering woman seafarer, who served as a ship's engineer
during the Second World War.

Thelma Furness (1904–70)
A twice-divorced American banking heiress and frequent transatlantic traveller, who often accompanied by her twin sister
Gloria Vanderbilt, Thelma was the long-standing mistress of
the Prince of Wales.

Martha Gellhorn (1908–98)
American-born writer and war correspondent with a sixty-year career and survivor of a marriage to Ernest Hemingway.

Hilda James (1904–82)
Olympic swimming champion from Liverpool who was employed by Cunard, first as a swimming coach and later as a social hostess.

Violet Jessop (1887–1971)
To support her family, naïve and unworldly Violet went to sea as a transatlantic stewardess aged twenty-one, and rapidly learned a great deal about human nature.

Nin Kilburn
Born into a Liverpool sea-going family, Nin lost a sister who was working on the *Lusitania* when it was sunk in 1915. Unable to find work as a schoolteacher in the 1930s, Nin went to sea, and achieved rapid promotion.

Hedy Lamarr (1914–2000)
Viennese-born actress who escaped a repressive marriage to a pro-Nazi Austrian arms dealer, and secured herself a future as a Hollywood star during a transatlantic voyage.

Mary Anne MacLeod (1912–2000)
The youngest child of ten born to a Scottish crofter, Mary Anne escaped her impoverished and depressed home island of Lewis by buying a one-way third-class ticket on a ship sailing to New York, seeking work as a domestic servant.

Maida Nixson (b?–1954)

Former journalist, writer, bank clerk and soft toy designer who became a stewardess in 1937 out of desperation and, to her surprise, loved the job. Her highly readable and enjoyable book, *Ring Twice for the Stewardess*, was published in the same year as her death.

Marie Riffelmacher

Fifteen years old in 1923, Marie was an economic migrant from Altenberg who escaped the turmoil of 1920s post-war Germany by embarking on a ship for America.

Edith Sowerbutts (1896–1992)

A much travelled and feisty character who enjoyed being a conductress caring for unaccompanied women and children on the North Atlantic route between 1925 and 1931. She later became a stewardess for Cunard/White Star Line.

Prologue

Smart and snug in heather-toned tweeds, a British-born lady passenger of middle years and independent means is settled in a teak steamer-chair on the promenade deck of an ocean liner heading west across the Atlantic. There is a whiff of healthy ozone from the sea, and fellow travellers taking the air variously saunter or speed past, competing for laps, heading to the gym or to the swimming pool. The attentive deck steward proffers a steaming mug of *bouillon*. So effective against seasickness, but, in order to be sure, she'll take another Mothersill's tablet. Just as that kind stewardess had predicted, the heaving swell of the Atlantic Ocean once they passed Ireland had proved to be a little too lively for comfort. Yesterday had crawled by in a disorientating, low-lit blur; she had spent hours wedged into her bunk, lying prone, wishing for death, while keeping an eye on the discreetly positioned *vase de nuit*, just in case. But this morning she had woken with the appetite of a Bootle-born coal stoker.

She had rung the bell (twice for the stewardess; a single ring would summon the steward, which would breach etiquette), and requested a fully-laden breakfast tray of kippers, tea and toast. Her equilibrium has now been restored in every sense; standing with her feet at the 'ten to two' ballet position for maximum stability, her knees slightly bent, as the stewardess tactfully suggested, also seems to help.

Bathed, dressed and ready to face the world, our heroine

has ventured out onto the covered deck and secured a reclining chair. Her thoughts turn to her fellow passengers: she must trawl through the alphabetically arranged list of first-class passengers' names left prominently in her cabin to see if she knows anyone else on board, or can spot anyone she might like to meet. 'Society' is superior to 'variety', and she is hoping to make the acquaintance of the elite also travelling in first class. In particular, she is looking out for a pleasant, well-heeled bachelor, as there is rather a dearth of those at home, just a few years after the end of the Great War. Perhaps she'll do better in America, Land of Opportunity.

Opportunities abound for sociable, single, wealthy, trans-atlantic female passengers; according to the daily newspaper printed every night on board, there is a tea dance this after-noon, with music provided by the pianist, and, after dinner, a concert, followed by dancing to the ship's orchestra. Daytime visits to the swimming pool with tutoring from the female coach, the Turkish baths where she is pampered by the masseuse, and restorative sessions with the lady hair-dresser promote a sense of wellbeing. In fact, the whole vessel seems to have been deliberately designed to appeal to lady travellers of all stripes. Reassuringly, the grand public rooms recall the better sort of country house. Every room is decked out in a different historical style, as though it had grown gradually, over the centuries, with each generation adding another wing according to the fashion of the day. Odd when you think about it, because this enormous ship was constructed by swarms of men in cloth caps, riveting together great steel plates in somewhere 'industrial', like Clydebank. With its Palladian-style lounge and Carolean smoking room, the ship's interior recalls the smart new hotels now springing up in European and American capitals. No wonder some wag referred to this ship as 'the Ritzonia'.

Meanwhile the stewardess makes final checks of the state-rooms of 'her' ladies before the captain's daily inspection at 11 a.m. Scrupulously clean and neat in her grey uniform, white cap and apron, she is a 'company widow', whose husband died when the *Lusitania* was torpedoed in 1915. To provide for her two small children following the Armistice she went to sea as a 'floating chambermaid', leaving the boys to be brought up by their maternal grandmother. Every fortnight they have a scant forty-eight hours together, when her ship docks in Southampton, her home port. She brings them American cigarette cards for their collection, and picture postcards of the Statue of Liberty. Her wages are supplemented by the tips she often receives from grateful passengers. She has been on duty since 6.30 a.m., and will fall into her bunk in the tiny windowless cabin she shares with another stewardess around 10 p.m., with throbbing feet

In the second-class smoking room, concentration is fierce as four American buyers play bridge. Three of them are middle-aged women; it is a career that suits entrepreneurial types with a head for commerce. Each has commissions from competing Stateside department stores or specialist boutiques, selecting European garments and accessories for the American market. They are returning from the summer shows at the Paris fashion houses with their precious purchases, samples and order books safely stowed in their cabins. Buying is a competitive business, and economic espionage is rife, so each of them will type up their orders themselves, rather than booking the ship's stenographer to do the work for them. One of the unspoken reasons for these bridge marathons is so that they all know where their rivals are during waking hours.

The ship's stenographer is busy anyway, having been engaged by an eminent lady writer to take dictation in her cabin. For any young woman with an outgoing personality, shorthand and a portable typewriter provide a ticket to travel the world in these boom years just after the Great War. Today's client is a well-known British authority on theosophy, and has been invited on a lecture tour of America; while her writings may be spiritual in tone, she is a canny business-woman with an eye to the benefits of meeting her public and, as one of her contemporaries, P.G. Wodehouse observed, 'She was half way across the Atlantic with a complete itinerary booked, before ninety per cent of the poets and philosophers had finished sorting out their clean collars and getting their photographs taken for the passport.'[1]

In another cabin further along the second-class corridor, an apparently respectable husband and wife are discussing their tactics for compromising a wealthy Frenchman, who is travelling alone. The couple are frequent transatlantic travellers, because they are professional blackmailers. On this voyage the wife has piqued their target's interest by convincingly playing the role of a bored and neglected beauty. She has hinted that her complaisant and lacklustre husband prefers to spend his nights afloat playing cards in the bar, and that she is lonely and available for a nocturnal dalliance. Tonight will provide an opportunity to reel him in, but first they need to divine his home address and the name of his wife. When, inevitably, the apparently outraged husband catches them *in flagrante delicto*, he will threaten to inform the Frenchman's wife unless the would-be philanderer agrees to pay 'compensation'.

Down near the waterline, in third class, professional chaperones provide support and protection for unaccompanied women and children from far-flung parts of Europe. This morning the chaperone is organising hot seawater baths for some of her reluctant charges, a welcome opportunity for them to bathe in privacy and to wash their clothes. All the women travelling in third class are economic migrants, and the luckier ones are making the voyage with their families, neighbours or friends. Some already have a 'stepping stone', relatives who can help them get established on the new continent. The teenage daughter of a crofter from the Scottish island of Lewis, with older sisters already settled in New York, is eminently eligible for a visa to work as a domestic servant. Two brothers and a sister have secured employment on a farm in rural Michigan; their wages will eventually enable their hard-pressed family back home to escape the economic ruin of 1920s Germany.

On this summer morning the oil-fired ocean liner is heading west at an impressive rate, covering 500 miles a day, steaming from the Old World to the New. These 2,500 souls from all over Europe – passengers of many classes, creeds and countries, and the 'ship's company': all the crew and staff, male and female – are under the omnipotent command of the captain while they are under way. Though it is one of the largest man-made moving objects on the planet, the liner is dwarfed by the ocean, and is totally alone on a vast and open sea. Anyone standing high enough on the very top deck, and rotating 360 degrees, would see nothing more than a slightly curving panorama of completely empty horizon, a daunting prospect for the agoraphobic.

But for the women on board the ocean liner, the great ship offers hope, opportunity, romance. Whether they are travelling for leisure or pleasure, by virtue of their celebrity or to preserve their anonymity, as matrons, migrants or millionairesses, as passengers or staff, the journey they undertake will change their lives for ever.

Introduction: Cresting the Waves

Remarkable as it seems nowadays, until the middle of the twentieth century the only practical way for civilians to cross the Atlantic was by embarking on some sort of ship. Three thousand miles of open ocean could not be traversed except by sailing, until technological and engineering advances made long-haul flights first possible, then comfortable and affordable for paying customers in the 1950s. Consequently, until the middle of the twentieth century, ocean liners were the only viable form of transport for international travellers between the landmasses of America and Europe.

For more than a century before flying replaced sea travel, mid-Victorian entrepreneurs, engineers, designers and ship-builders collaborated to create technologically advanced great ships, which were built to sail the Atlantic – a daunting stretch of water, prone to sudden storms, thick fogs and occasional icebergs. The ships were primarily designed to carry raw materials, saleable commodities and produce, and important communications such as mail on a regular and reliable basis. They were operated for the maximum profit of the privately owned shipping companies who competed for the most lucrative routes between the major ports of Europe and those of the American eastern seaboard, and as a sideline they also transported people who were prepared to face lengthy, uncomfortable and often dangerous journeys.

Each ship had to be entirely self-sufficient while at sea,

carrying its own provisions, including fuel, food and fresh water, labour, tools and expertise, as it would be out of sight of land and far from any form of assistance for weeks on end. In addition, the vessel had to be resilient enough to withstand the rough seas and adverse conditions to be found in the North Atlantic. Engineering innovations such as the steam engine and steel hull enabled the construction of ever-larger and faster ships, ramping up competition between world powers of the time, especially between Great Britain and Germany. Each sea-going nation competed to advance the interests of its own mercantile fleet, meeting the astonishing demand for transatlantic travel.

In January 1842 Charles Dickens sailed from Liverpool to Boston on RMS *Britannia* with his wife Kate and her maid. He was impressed by the ship's power and speed – revolutionary for the time – but was appalled by his cabin, which was 'utterly impracticable, thoroughly hopeless and profoundly preposterous too … Nothing smaller for sleeping in was ever made, except coffins,' he wrote lugubriously.

Dickens kept a vivid account of their appalling eighteen-day journey, describing the ever-present seasickness, the tumultuous roughness of the sea, the perpetual discomfort of being cold and wet, the difficulty of dispensing remedial teaspoons of brandy as the ship heaved and plunged, and the sorry plight of the lone cow, brought along to provide fresh milk during the passage, lowing piteously in her stall up on deck. The *Britannia* was badly damaged by ferocious storms, one of the lifeboats was smashed to matchwood by a freak wave, and Dickens doggedly and soggily played cards hour after hour with the ship's doctor for fifteen days.

Once the worst of his nausea had receded, Dickens mentions, almost in passing, that a stewardess provided meals for the passengers in the communal saloon: 'a steaming dish of baked potatoes, and another of roasted apples; and plates of pig's faces, cold ham, salt beef; or perhaps a steaming mass of hot collops ... not forgetting the roast pig, to be taken medicinally'.[1] This appears to be the first mention of a stewardess in the literature of the era. Dickens appreciated her diplomacy and positive attitude in dealing with him and his suffering womenfolk:

> God bless the stewardess for her piously fraudulent account of January voyages! ... and for her predictions of fair winds and fine weather (all wrong or I shouldn't be half so fond of her); and for the ten thousand small fragments of genuine womanly tact [in reassuring passengers] that what seemed to the uninitiated a serious journey was, to those who were in the secret, a mere frolic, to be sung about and whistled at![2]

Dickens was sanguine about having a female crew member dispensing wisdom and good cheer. While women seafarers were not common, neither were they a complete anomaly in the first years of the Victorian era. Their occasional presence on board ships started informally, but by 1840, Union Line, Royal Mail Steam Packet Company, and Peninsular and Oriental Line (P&O) regularly employed women to sail on their ships to destinations as far away as Australia, India, South America and the West Indies. In previous centuries the only female acceptable on board the more traditional ships had been the carved figurehead on the prow, usually depicting a voluptuous bare-breasted woman, believed to bring good luck. Although mariners of all types have always referred to their ship in the feminine, as 'she' rather than 'it', paradoxically the superstitious belief persisted that it was unlucky to carry women at sea. However, by the beginning of the nineteenth century wives occasionally accompanied their husbands on naval warships. In time, whaling ships and smaller merchant vessels often carried the captain's wife, especially for lengthy voyages. Some British women actually worked on ships, particularly those who came from seafaring families. On board they had considerable status as they were the boss's relative, and would oversee the catering, provide nursing care, and deal with the accounts by bookkeeping, paying wages and provisioning. In the larger ships they mingled only with the senior officers and tended to stay close to the captain's quarters, to maintain some distance from the rougher crewmen.

By contrast, some intrepid women went to sea in order to become sailors themselves, and disguised themselves as youths so they could join merchants' ships as cabin boys, where they were less likely to face scrutiny, exposure or censure than in the Royal Navy. In 1859 *The Times* newspaper reported a curious case of 'A Female Sailor', an unnamed woman who was charged at the Newport police office in Monmouthshire with wearing seaman's clothes. It emerged that for ten years the woman, accompanied by her husband, had 'scorned her proper clothes' and had travelled the world working as a sailor, loading and unloading cargoes with the rest of the crew, without her true gender being revealed until her secret was accidentally discovered on arrival at the docks of Newport. The story was related in a tone of some puzzlement; a cross-dressing married woman who succeeded in passing herself off as a hard-working sailor for a decade transgressed every social boundary.[3]

In Victorian Britain, seaports were a vital part of the industrial economy. They were the gateway through which raw materials from all over the world were imported, and the route by which manufactured goods churned out by the burgeoning factories and mills were taken overseas for sale. The docks of Liverpool, Bristol, Cardiff, the City of London, Southampton, Glasgow and Harwich were teeming with ships of all sorts, from vessels exporting coal and importing cotton and timber to mail boats and passenger ships, whose main cargo was people.

Women's lives were tied to the shipping trade long before they started to travel and work on the ships. For shipowning families, their livelihood might be invested in a single ship, captained by a family member and manned by local seamen, or they might own a number of vessels and employ crew for specific commissions, carrying cargo and passengers around the world. In smaller, family-run firms, the women often played an active part in the business; acting as clerks, they attended to business matters while the men were away. Ports such as Portsmouth or Plymouth had ample opportunities for women to make a living, from running inns catering for travellers, or lodging houses for the crew on shore, to owning chandlers' stores and selling provisions to shipowners. They might work as seamstresses, sailmakers, laundresses or cooks, or own a small shop. Those were the more respectable port professions, but there was also the flourishing business of prostitution, as there was always a ready market for transactional sex in any port. In addition, there were the pubs and the gambling dens, where a recently returned sailor could soon be relieved of a large amount of his pay, and rich pickings existed for pickpockets of both genders.

By the 1880s, as emigration to the New World grew, passenger ships increasingly carried women and children as well as men, and in response the shipping companies took on female crew to attend to their specific needs. In addition, ships were designed and marketed to appeal to potential female passengers. By the last decades of the nineteenth century, forward-looking companies such as Cunard stressed their vessels' comfortable facilities and underlined the emphasis on safety, important considerations when a passage to America from Europe could take weeks and was often memorable for unprecedented levels of physical misery. The vast majority of Europeans travelling west across the Atlantic in the latter half of the nineteenth century were relocating permanently to North America. Between 1860 and 1900, 14 million people emigrated from Europe to the United States, of whom 4.5 million travelled from Liverpool, half of them on Cunard ships. The founder of the line was Samuel Cunard, a Canadian businessman who had won the first British contract to deliver mail by ship across the Atlantic. His new vessel, a no-frills paddle-steamer, RMS *Britannia*, was primarily intended to transport mail and cargo, but it also had accommodation for 115 passengers. On its inaugural voyage on 4 July 1840, Samuel Cunard was accompanied by his daughter, so confident was he that the ship was safe.

When merchant vessels like RMS *Britannia* began carrying female passengers, it was considered desirable to have a woman crew member on board, so that 'proprieties could be observed', and so women were afforded new opportunities in the maritime trade – opportunities that allowed them to travel. They acted as chaperones, ministering to the passenger if she was unwell, ensuring her privacy and dealing with all the personal hygiene issues likely to arise on an ocean-going trip lasting many weeks. While the crew might have private misgivings about having women on board, they were overruled by the captain or master, especially if he was keen to have his own wife along on the trip.

As the passenger trade grew in the late Victorian era, some shipping companies actively recruited small numbers of British women to work at sea, predominantly as domestic employees on passenger ships. The Merchant Shipping Act of 1875 required all passenger ships carrying migrants of both sexes to carry on board a matron, who would look after the interests of female and child migrants in steerage, ensuring they were kept segregated from male passengers if travelling unaccompanied, and that they were not importuned by the crew. On the upper decks, too, stewardesses fulfilled the combined roles of chambermaids, personal maids and sometimes nurses, tending to female passengers in physically uncomfortable and cramped conditions. At first, stewardesses were the wives of ship's stewards and travelled merely as their husbands' assistants. By the 1890s they were increasingly recruited from the ranks of 'company widows', married women whose deceased spouses had been sailors or ships' officers.

In an era before any form of social welfare, widows were often left facing destitution, and they needed to find respectable, regularly paid employment to maintain their families. Stewardesses often had young children, who would be left in the care of relatives or friends while the mother went to sea to earn an income to support her offspring. Women with very young offspring preferred working the short-haul trips across the English Channel, though it was far more lucrative to sign on for a lengthy ocean trip lasting weeks, sailing from Liverpool or Southampton, to North or South America. Basic salaries were modest: a stewardess working for Royal Mail in 1879 was paid £3 a month, the modern equivalent of less than £50 a week. However, there were tips to be earned from passengers for good service, and at least the stewardess had bed and board provided while afloat, which was a consideration in households where money was tight.

Stewardesses of the Victorian era were distinctively if repressively dressed in dour uniforms which suggested the garb of prison warders, nuns or nurses. Victorian society was rigidly segregated by gender at all levels and throughout all classes, and life on board ship reflected these social mores. It was generally believed that a decent woman's place was in the home; any respectable female should aspire to be a 'domestic angel', maintaining a household, caring for a family. The culture on board ships tended to be ruggedly masculine, and most of the men aboard were completely unused to females in the workplace. The Captain's Lady had a certain status because her husband was the boss, and it was understood that she travelled under his protection. Nevertheless, ship's officers and crew members tended to find it hard to know how to deal with women as fellow workers. Were they 'ladies', who should be regarded as the fairer sex, or were they 'shipmates' in skirts? Where were the boundaries? Many sailors, although usually married and having families (sometimes several simultaneously, in different ports), had only limited contact with working women throughout their adult lives, encountering them primarily as barmaids, landladies, shopkeepers, cooks, brothel-keepers or prostitutes. Those women who wanted to work at sea also risked their reputations ashore, because they might be assumed to be soliciting for sex, as prostitution was the female trade most often associated with the gritty, commercial reality of ports, harbours, docks, pubs and sailors.

As a result, seafaring men's initial reactions to having

female shipmates working alongside them ranged from dubious mutterings and muted disapproval to outright bullying or sexual harassment. If the women were lucky, a certain amount of creaky gallantry might be expended on them, though this also could be disadvantageous. Stewardesses had to be careful in their dealings with the crew and male officers on board, maintaining a fine line between personal reticence and affability, and had to be adroit at sidestepping unwanted male interest. They should be neither too friendly, nor too distant. They were not allowed in male quarters, neither could they work alone in a cabin with a man, nor attend to a solo male passenger. While not on duty, the female crew were expected to keep to their cabins, in case their mere presence on board inflamed male ardour. To circumvent this potential clash of the sexes, the shipping companies preferred to recruit older women, who were determined to maintain their dignity. In addition, male and female crew were discouraged from fraternising socially while off duty, although friendships among shipmates inevitably developed, and, as one former stewardess fondly recalled if she was able to totter off a ship on arrival at port, she could always find a congenial male shipmate with whom she could explore the city.

In the mid-nineteenth century very few women formally worked at sea; according to Dr Jo Stanley, in Southampton, out of 4,500 local inhabitants listed on crew logs between 1866 and 1871, only twenty were women. However, the numbers of female passengers increased greatly towards the end of the century, not least because the means of travelling became less arduous and less dangerous. Women of all nationalities were crossing the Atlantic in both directions out of necessity or choice – to emigrate, to join their families, to find work in the New World or a husband in the old one. It was recognised by the major shipping firms that socially superior female passengers should be specifically catered for; Cunard introduced the first lounge exclusively for women on SS *Bothnia* as early as 1874. This ship was also the first to have a system of electric bells in the first-class cabins to summon assistance, an innovation much appreciated by any passenger laid low by seasickness.

By the 1890s there was a boom in passenger shipping, and the major ports of Britain became the embarkation points for emigrants from all over Europe as well as the British Isles. In 1893 large transatlantic liners started to leave from Southampton as well as from Liverpool, which had previously been the main embarkation port for North America. The White Star and Cunard shipping lines joined P&O in setting up offices and infrastructure in Southampton, which was just an hour from London by train. With the growth in the number of female passengers, gender-specific roles such as bathing attendants, nursery nurses, laundry attendants and masseuses were also created aboard the big ships. The shipping companies received applications for women's roles on board the ocean-going ships far beyond the number of positions available. The jobs were physically demanding, and those taking them would be living away from home – often in cramped, communal quarters – but, in an era of limited job opportunities for women, the idea of going to sea and earning an independent living was appealing to many.

As the new century dawned, wealthy women passengers required experienced, discreet, knowledgeable female attendants to assist them with the everyday business of sea travel. Even if a lady was accompanied on a voyage by her husband, members of her family and her maid, she still needed a stewardess, whose role was to minister, to reassure and to provide practical assistance to all of them. No respectable woman, travelling alone or with her spouse, would want a male steward to enter her cabin to bring her food, change her bedding, deal with the slops and chamber pots, or care for her when she was ill or seasick. Neither should a lowly steward catch sight of an upper-class woman in *déshabillé*, or looking anything other than perfectly groomed, poised and fully-clothed. In an era when getting the mistress into or out of her clothes could take twenty minutes of a competent maid's time on dry land, it can be imagined how much more difficult the task was in the confines of a pitching, rolling and ill-lit ship's cabin during a winter storm. In addition, lady's maids were not immune to seasickness, while experienced stewardesses tended to be more resilient and experienced, adept at solving the everyday problems on board that seemed unfamiliar and even alarming to the passengers. Consequently, there was a growing demand for stewardesses to meet the surge in transatlantic female travellers, although the limited positions available were not easy to secure. As a careers book for women from 1894 discouragingly described it:

> One would imagine that there were not many women who would care to occupy the post of stewardess on board an ocean liner, yet when a vacancy occurs there is never a lack of applicants. In fact, every steampacket company has a long list of names from whom a choice is made when necessary ... the work is hard and disagreeable; it demands untiring energy, and a temper which nothing can ruffle; it shuts one off from home and from home comforts.

Besides possessing some knowledge of nursing, a stewardess must be proof against sea-sickness, even in the worst weather, and she must at the same time be able to sympathise with those who in this respect are weaker than herself. This is a combination not often met with. No; a stewardess is certainly not overpaid.[4]

It is easy to dismiss stewardesses as merely 'floating chambermaids', ocean-going servants, but in fact their expertise, their 'sea knowledge' and practical skills made them far less subservient than their domestic equivalents. They saw themselves as ladies who assisted their passengers, rather than serving them. A stewardess's interactions with each individual passenger lasted only for the length of that voyage and were transactional in nature. She reported to the chief steward, who in turn reported to the purser, and she was employed by the shipping company to provide care and support to a specified number of passengers while they were in transit.

By the early twentieth century, the liners ploughing between Europe and North America were the largest man-made objects ever created. Wrought from mighty steel components, riveted together in shipyards by tens of thousands of skilled workers, and powered by engines that consumed hundreds of tons of coal per day, these behemoths were at the cutting edge of industrial technology. Engineering expertise was dedicated to carrying hosts of people thousands of miles, year after year.

The ship itself was a complex, technologically advanced, floating city in microcosm, virtually self-sufficient in terms of supplies, victuals, fuel and skilled labour. One seafaring officer around this time referred to his ship, RMS *Aquitania*, as 'the steel beehive', because of the way it constantly hummed with creative activity, purpose and a sense of common enterprise. Each vessel had been designed and engineered specifically to cross one of the world's widest and least predictable expanses of water every week, in order to deposit its wide and varied human cargo safely on the American side, and to return with people wanting to explore Europe.

The captain was the ultimate authority at sea, and he made the crucial decisions about the vessel, its passengers, crew, cargo and course. It was an extremely responsible job, as the fate of hundreds, if not thousands, of people on board were dependent upon his decisions. To qualify, a man needed decades of experience at sea, working his way through the strict and exacting merchant navy hierarchy. 'Master before God' was the impressive term written on his ticket, his 'licence' to captain a ship. Many captains were extrovert personalities, and some cultivated their personal foibles or larger-than-life characteristics, which added to their popularity among passengers and ensured repeat custom. To be invited to dine at the captain's table was a great honour for distinguished passengers.

There were three main departments on board, each of which reported to the captain. The deck officers' duties encompassed safe navigation, manoeuvring, steering and docking of the ship, and care of the mail, baggage and cargo. The engineering department oversaw the performance and maintenance of the ship as a complex, moving vehicle. The victualling department was headed by the purser; he was the officer who had the most contact with the passengers of all classes, and he managed all the female staff who worked on board the ship, as well as the catering staff. He needed numerous assistants, such as stenographers, to fulfil administrative roles, as well as stewards and stewardesses, swimming pool attendants, barmen, hairdressers, bath stewards, bellboys, barbers and the band. In reality, most of these roles were managed by the purser's deputy, the chief steward. Passengers who 'knew the ropes' would butter up the purser, as he was the ultimate arbiter, in charge of assigning cabins, smoothing out problems, organising entertainment and excursions, changing money and storing valuables. Ambitious travellers hoping to get to know their wealthy celebrity fellow passengers would be grateful to the purser for ensuring proximity to the object of their attentions

Among the ship's company there was some resentment between the different roles. Deck officers thought of themselves as socially superior to the rest of the crew, and were instructed to mingle with the first-class passengers and to be charming. They often looked down on the engineers, dismissing them as 'grease monkeys', but the engineers saw themselves as the real talent on board, the masters of their complex, sophisticated ocean-going machine. But some of the friction between the different departments on board can also be attributed to clashing assumptions about masculine and feminine roles. The more rugged elements of the ship's company could be disparaging at the thought of a steward 'fawning' over his passengers while waiting at tables, or tucking them up in a sheltered deckchair with a warm rug over the knees. Making beds, tidying cabins, and pressing evening clothes were not deemed proper men's work by the officers or engineers. By contrast, the care-giving and nurturing roles traditionally associated with women were gradually accepted by their male colleagues as having a place on board passenger ships. Nurses and stewardesses were seen by their more enlightened male shipmates as 'ministering angels', so long as they confined their activities to looking after their charges.

The structure of the early twentieth-century ship reflected the social hierarchy of the Edwardian era, divided laterally by strata, like geological deposits. Indeed, strenuous efforts were made to keep the more prosperous passengers totally unaware of the conditions in which the less well-off were travelling. A single ship could be carrying millionaires, moguls and monarchs on its upper decks, and thoroughly respectable clerics, merchants and middle-class families in its second-class quarters. The passenger manifest would list the names and titles of those who were able to travel in some luxury. A traveller's experience of a transatlantic voyage was largely dependent on how much they paid for their tickets. Cunard's flagship, the *Mauretania*, boasted luxurious regal suites, each with two bedrooms, a dining room, a butler's pantry, a reception room and bathroom, with dramatic views of the ocean beyond the porthole. By contrast, cabin accommodation for third-class passengers hundreds of feet below in the same vessel consisted of a windowless cell containing two rows of upper and lower bunks, separated by a toilet seat placed over a bucket.

The uppermost layer of each grand ocean liner comprised the *gratin*, those passengers who could afford the luxurious staterooms, the self-contained and opulent suites on the highest deck. Many had their own balconies, private bathrooms and adjacent rooms for their maids and valets. First-class travellers dined in exclusive restaurants, barred to socially inferior interlopers. Directly below them were horizontal bands of more modest second-class cabins, well-designed and comfortable, for the respectable international travellers of genteel habits, who knew their way around a decent menu, which was provided in their segregated 'dining saloon'. Their cabins were more compact, with two bunk beds, but their aspirations for social advancement were high.

And then there were the hundreds of third-class passengers, located near the waterline, occupying the low-ceilinged, rather dimly lit decks above the cargo hold. So closely packed were the humble third-class customers that it was their numerous if modest one-way fares that made the whole voyage financially viable; despite appearances, the numerous poor supported the whole enterprise.

Before the Great War the vast majority of people travelling west from Europe to the North American continent were immigrants in search of a better life. They travelled in challenging and uncomfortable conditions, but they were only going to make this journey once, in one direction, and so were prepared for hardship. In the first two decades of the twentieth century, approximately 14.5 million immigrants arrived in the USA. It has been estimated that between 1830 and 1930 over 9 million emigrants sailed from Liverpool alone, heading for new lives in the US, Canada and Australia. For much of this period Liverpool was the most important port of departure for emigrants from Europe because, as well as its established transatlantic links, it could accommodate the many emigrants from the countries of north-western Europe, such as Scandinavians, Russians and Poles who crossed the North Sea to Hull by steamer, and then travelled west to Liverpool by train.

Edwardian third-class accommodation was basic though adequate, but its Victorian predecessor, known as steerage, had been much worse. The dormitory-style accommodation would start out in a reasonably hygienic state, but the dense occupancy, lack of ventilation and inevitable seasickness could make it hellish during the journey. Each occupant was allowed approximately 100 cubic feet of space. Each berth was six feet long and two feet wide, and comprised iron- or wooden-framed bunks. Thin mattresses were stuffed with straw or dried seaweed, and if a passenger hadn't brought their own pillow, they improvised with their life jacket. Only a single blanket was allowed per person, so to keep warm and to preserve one's modesty most people slept in all their clothes for the entire voyage. By the early twentieth century, third-class accommodation on the better lines generally provided a functional but adequate communal dining room, and passengers slept in four- or six-berth cabins.

For these third-class travellers, anxiety levels must have been high. Aside from the considerable physical discomforts, there were all the individual reasons people had left home in the first place: to avoid persecution, conscription, jail sentences, onerous obligations, creditors, betrayed lovers, abandoned spouses or old enemies. Hardened criminals rubbed shoulders with blameless farm labourers. Some passengers had previously travelled no more than ten miles from home, or as far as they could walk in a day. All those emigrating were aware they were taking a great gamble, investing their hopes in an unknown future in a totally alien country.

For European-born women who wished to emigrate to America, there were ample opportunities to find paid work, albeit mostly of a domestic nature. It was widely known in Europe that there was great demand in America at the time for servants, so many women arriving at New York stated their occupations to be 'domestic service'. However, this was probably a strategy merely to gain admittance, because, as an influential female journalist called Frances A. Kellor observed, 'opportunities are open to them to enter any trade, profession or home for which they fit themselves. The immigrant, then, is a transient, not a permanent, domestic worker.'[5]

Hordes of women made it across the ocean to the United States, and were subjected to the forensic scrutiny of the immigration officials at Ellis Island, whose job was to assess each applicant, to scrutinise their potential to be a useful and productive member of society, and to winnow out anyone deemed unworthy of becoming a new American. Immigrants who were unlucky enough to harbour identifiable diseases were refused entry; those who were too feeble in mind or in body would be sent back to their homelands on the ships that brought them.

Millions of women's lives were transformed by their journeys between the Old and New Worlds in the first half of the twentieth century. From the beginning of the First World War to the end of the Second, women of all cultures and backgrounds went to sea for an infinite variety of reasons. For some passengers a sea journey was a welcome and pleasant interlude in their well-padded lives, a week or so spent in a giant floating resort hotel, with all the genteel amusements and luxurious indulgences that money and deferential servants could provide. A new breed of independent women, those left single and self-supporting as a result of the appalling loss of men during the Great War, were also on the move between one continent and another, seeking careers in showbusiness, the arts and commerce. For them, the great ships were the Atlantic Ferry, linking their homelands with their career opportunities, and they were frequent travellers. Down near the waterline, there were women who staked their futures on a single voyage to the other side of the world – an experience that was daunting, both as an endurance test and as a brave leap of faith towards an unknown future.

And for a select band of adventurous working-class women the ship itself provided their living, often for decades; it was their workplace, their social life, their 'home from home'. Taking a job as a stewardess, a nurse, conductress, hairdresser or hostess on one of the great ships of the 1920s and 1930s enabled them to gain a degree of financial autonomy and independence unparalleled on land. Though small in number compared to their male counterparts, women seafarers had opportunities for social advancement, for making friends, for travelling the world and being amply rewarded for their labours.

Their individual stories, of camaraderie and calamities, provide fascinating and informative narratives about the changing nature and experience of transcontinental sea voyages, in an era of unprecedented, immense social and cultural change. *Maiden Voyages* is a celebration of the diverse journeys made by a number of intrepid heroines, drawn from many countries and different classes. This is a collection of selected biographical tales, both cautionary and life-affirming, about dynamic women on the move, set primarily between the two World Wars, during the golden age of transatlantic travel.

I

Floating Palaces and the 'Unsinkable' Violet Jessop

In 1908, at the age of twenty-one, Violet Jessop embarked upon her maiden voyage as a stewardess on the *Orinoco*, a steamer sailing to and from the West Indies carrying mail, cargo and passengers. It was her personal circumstances – a combination of family connections and financial desperation – that drove Violet to seek a job on board ships and it set her on a maritime career that spanned forty-two years and incorporated more than 200 ocean voyages.

It required a leap of faith to emigrate to an unknown land in South America in the Victorian era, but William Jessop was an incurable optimist. Born in Ireland, he sailed to Argentina in the mid-1880s, and his fiancée Katherine Kelly, from Dublin, joined him in 1886. They married and set up a sheep farm on the pampas; they were very much in love and prepared to endure their spartan living conditions in the hope of a better life. Their daughter Violet, the first of six children, was born in 1887. With a rapidly growing family and a dwindling future as a rancher, William Jessop gratefully accepted a better-paid post in Buenos Aires. Violet's child-hood years in Argentina were largely happy, though she contracted tuberculosis and after a spell in hospital was sent to recuperate in the clear mountain air of the high Andes. She was an intelligent child, a keen reader with a natural talent for languages, and she helped to nurse two of her

27

younger brothers through diphtheria. Then tragedy struck the Jessop family: Violet's beloved father William died during an operation, leaving Katherine and the children bereaved, and bereft of funds.

Mrs Jessop was advised that if they remained in Argentina her sons would be called up for military service, so she took all six children to England. The voyage across the South Atlantic was Violet's first experience of ocean travel. Once in England, Mrs Jessop managed to get work as a stewardess on a Royal Mail packet line, while Violet cared for her brothers and sisters at home. But in 1908 Katherine Jessop was forced to retire from the sea due to ill health, and so Violet abandoned her plans to become a governess and applied to be a stewardess with the same company, to provide a modest income for a large and otherwise unsupported family.

Vivacious and slim, with good clothes sense, dark hair, grey eyes and long eyelashes, Violet had a lilting, slightly Americanised Irish brogue. The victualling superintendent who interviewed Violet was initially inclined to turn her down on the grounds that she was too young and attractive – most shipping companies tended to recruit only respectable widows of mature years, favouring those whose husbands had died while working for the company. But Violet promised to be 'most circumspect and careful if he gave me a post as a stewardess',[1] and so he relented, and she became a stewardess on board the *Orinoco*.

Despite her grey, dowdy and old-fashioned uniform, evidently designed to make its wearer as unattractive as possible, Violet attracted a great deal of male attention in her new role. But her transparent naivety and innocence were strangely protective as she did not initially recognise the motive. 'Though aware of the many efforts being made to help me, I did not realise at the time that youth, feminine youth, is almost a fetish to seafaring men and has a tremendous power over them, exacting willing service from uncouth and often uncivil men. I was not to know till years later that the adulation I had accepted as chivalry was largely a demonstration of sexual attraction.'[2] Most of her fellow seawomen were middle-aged or even elderly – less likely to inflame the ardour of the male crew, and more adept at sidestepping any unwanted attentions from on-board Lotharios. Double standards applied too: because Violet had spurned his advances, one captain dismissed her for 'flirting with the officers'. For women seafarers there was always a fine line between general conviviality and appearing to invite intimacy. It was extremely difficult to keep any romantic spark secret anyway; passengers and crew were trapped in close proximity, and as one day merged into another, any meaningful glance in a galley or shared moment on the crew deck, no matter how innocent, inevitably sparked speculation and gossip among one's fellow shipmates.

There was a hierarchy among the stewards and stewardesses; everyone wanted to work in the first- or second-class sections of the ship, since 'perks' from the better-off clientele were a vital part of the job, but in third class tips would be almost non-existent. Violet observed that the stewards often sniped at each other, unable to express their frustration at the passengers who harried them every hour of the day. Tips were handed over at the end of the voyage, rather than for individual services as and when they were provided – the normal practice in hotels – so the stewards and stewardesses had to maintain high standards of service throughout the voyage. With each day that passed, the pressure on them to keep their charges happy increased.

Violet shared a tiny cabin on the *Orinoco* with another stewardess, who was a vinegary and forbidding character, and she battled seasickness and exhaustion in equal measure until she acquired her 'sea legs'. The basic pay and conditions were poor: she earned £2 10 shillings a month, two-thirds of her male colleagues' pay of £3 15 shillings. Stewardesses worked sixteen hours a day for seven days a week while afloat, as did the stewards, and the workload was immense – a constant round of answering cabin bells, washing floors, making beds, emptying slops and providing meals and drinks to female passengers. Violet was constantly either nauseous or ravenously hungry, sometimes simultaneously, and lived off tepid snacks consumed while standing in her steamy pantry, surrounded by the detritus of other people's leftovers.

Despite the drawbacks, the long hours, physical discomforts and frequent challenges, Violet endured the sea-going life for the sake of her earnings, which supported her mother and younger siblings back in Britain. While her salary was basic, she gleaned many tips, especially once she began working on the more lucrative transatlantic passenger routes run by White Star Line between Britain and America. It was her eventful career with this company, owned by J. Pierpoint Morgan's IMM, that earned her the soubriquet of 'The Unsinkable Stewardess'.

In 1911 Violet joined RMS *Olympic*, the largest civilian ship afloat at the time, and the flagship of White Star Line. It had been built by Harland and Wolff in Belfast, the first of three colossal 'sister' vessels to be completed. Violet was aboard on 20 September 1911, when the *Olympic* left Southampton and accidentally collided in the Solent with a British warship, HMS *Hawke*. There was significant damage to both vessels, including two large holes torn in the *Olympic*'s hull, but no fatalities, though it took eight weeks to repair the damage.

Violet's familiarity with the *Olympic* suited her for the maiden voyage of the liner's prestigious sister ship the following year. Aged twenty-four, she was engaged as one of the twenty-three stewardesses working with 322 male stewards on board RMS *Titanic*, which was to set out from Southampton to New York in April 1912. The *Titanic* boasted the ultimate in new technology, as well as unparalleled levels of luxury; Bruce Ismay, the chairman of White Star Line, described it as 'the latest thing in the art of shipbuilding; absolutely no money was spared in her construction'.[3]

But this impression of ease and confidence was to be shattered on the fourth evening of the *Titanic*'s maiden voyage, when the ship took evasive action too late to avoid hitting a massive iceberg. In the hours that followed, 1,517 people died. The disaster was due to a variety of factors, some of them quite mundane. The key to the locker containing binoculars, essential for those watching for hazards at sea, had been lost, so the lookout didn't spot the vast iceberg till it was too late. The two wireless operators on board had received at least six messages from other ships in the area, warning of ice ahead. However, they were employed by the Marconi Company, not by White Star, and therefore they were not seen as part of the ship's crew. Their incoming messages were relayed to the bridge, but appear to have been ignored by Captain Smith, who ordered the ship to proceed westwards at full speed, so great was the pressure from Bruce Ismay, who was on board, to get to America in a record-beating time. As a result, the great ship collided with a huge iceberg, lit only by starlight. Once the collision had gouged a 300-foot-long underwater gash along the side of the hull, a fundamental design flaw allowed seawater to pour from one unsealed compartment into another, causing the ship to list and precipitating its demise.

At first the passengers were reassured by the ship's company that the ship was unsinkable, and encouraged to return to their cabins. That misconception caused a fatal delay for many, exacerbated by ill-discipline and incompetence in filling and launching the lifeboats. As the ship listed it became increasingly difficult to assist passengers into the boats, and to launch them successfully. So confident had the designers been that this magnificent vessel could not sink that the *Titanic*'s lifeboats could only carry about 52 per cent of the 2,207 people on board, and most of them were launched only half-full. The people who perished in the greatest numbers were those lower down in the ship, the steerage passengers who were denied access to the upper decks even as the ship's fate became evident. And, of course, the crew, many of whom stayed behind to keep the ship's engines, lighting and communications systems going as long as possible. Most men on board insisted 'women and children first', from a sense of gallantry, but often wives refused to leave without their husbands. However, not all the men on board were so noble: Bruce Ismay escaped in a lifeboat and survived, though he was subsequently to live the life of a recluse after a nervous breakdown.

Violet Jessop and her friend, another stewardess called Ann Turnbull, had retired to their shared cabin below *Titanic*'s waterline. They were woken at 11.40 p.m. by a 'rending, crunching, ripping sound', which lasted about eight seconds. The stewardesses dressed quickly and went to their respective cabin sections to wake up women and children, and to make sure they had life jackets. They climbed to the boat deck, and Violet went to her assigned lifeboat, number 16, where she was handed an abandoned baby, which she wrapped in a quilt. Her survival was due to a young officer urging her to clamber into Lifeboat 16, to demonstrate to reluctant passengers who didn't speak English how it was done. Her example was followed, and the boat was quickly lowered towards the water with its thirty occupants: twenty-three women, five crewmen, Violet's infant charge and a small Lebanese boy. The oarsmen quickly struck out across the cold, still waters to get some distance from the sinking ship, as they were afraid of being dragged underwater if the sea rushed in to fill the huge voids of the ship's funnels. Over the coming hours they watched in horror as the disaster unfolded and the sea filled with a litter of desperate, drowning people, life jackets, overturned and damaged boats, rafts, deckchairs and all the floating impedimenta associated with a luxury liner. After the *Titanic* slipped below the surface, its rows of lights extinguished as they were swamped by water, those in the bobbing lifeboats were left floating in the cold and dark, listening to the diminishing cries of others pleading for help as they succumbed to hypothermia.

The following morning, near dawn, Violet and her fellow survivors saw a Cunard steamer, the *Carpathia*, approaching carefully through the ice. It had been heading for the Mediterranean when it had picked up the distress calls from the *Titanic*, despatched by the Marconi men, and had raced at full speed to the stricken liner's last known position. The crew suspended Jacob's ladders from the *Carpathia*'s port doors, and those in the lifeboats with sufficient energy climbed up. Small, traumatised children were winched aboard in mail sacks or ash bags – Violet remembered they looked like kittens. Violet, still clutching the baby, was hauled up in a bosun's chair. The crewmen who rescued her had to unfold her arms gently in order to release the infant from her grip, as she was so cold. By chance, the baby's mother had also been rescued and was already on board the *Carpathia*; she quickly reclaimed the infant and bore it away, without a word of thanks to Violet.

Maud Slocombe, the *Titanic*'s Turkish bath attendant, had also survived the sinking and was brought aboard the *Carpathia*. Her recollections of the collision were that she had hurriedly thrown on a coat over her nightdress when ordered to 'get up on deck' by one of the stewards. For an hour she awaited further instructions, as most of the officers refused to acknowledge that the ship might sink. Finally, she was put into Lifeboat 11, in charge of a baby, with seventy-two people, two of whom were stewards. This was the last boat successfully launched, and as the oarsmen pulled away across the water, the occupants listened in dismay and pity to the eight musicians left on board playing 'Nearer My God to Thee', before the *Titanic* sank.

In later years Maud remembered that the Italian doctor on board the *Carpathia* organised brandy and warm blankets for those rescued, but there was hardly any room for the hundreds of numb survivors. The passengers were generous, giving up their cabins to those craving rest and privacy, and donating their spare clothes to anyone who had escaped death in whatever they had been wearing when they clambered or were thrust into a lifeboat. Some women were still dressed in their fine silk or velvet evening gowns; others were in nightwear. The grim mood among the survivors ranged from unassuageable grief to outright hostility. Some of them expressed outrage that 'these common women of the crew', such as Maud and Violet, had been saved, while their own husbands had been lost.

When the *Carpathia* reached New York, White Star officers boarded and forbade any crew from talking to the press, who were clamouring to interview them. There were frenzied scenes on the dockside as relatives and friends of those known to have boarded the *Titanic* searched for familiar faces among the stricken survivors. The remaining crew were returned to Britain on the SS *Lapland*. A photograph shows thirteen female *Titanic* crew members standing, facing the camera, on the quayside at Plymouth. The third from the left is Violet Jessop, and the fifth is Maud Slocombe. They were taken on to Southampton, their home port, only to find that their pay had been stopped by their employers the night the *Titanic* sank. Needing the money for family commitments, the survivors could not afford the luxury of recuperating from their experiences. Within three weeks of the sinking of the *Titanic*, Violet was once again on the transatlantic run, working on the *Olympic*, and shortly after Maud took a posting on a White Star Line cruise to the Mediterranean.

To return to working and living on vast ocean-going ships so soon after surviving the trauma of the sinking of the *Titanic* doubtless required considerable bravery and resolve. Both Violet and Maud were aware that respectable and reasonably well-paid careers for working women at that time were limited, and that there would be others competing to take their jobs if they refused to return to sea. Women seafarers' personal circumstances often provided compelling motives for their willingness to face the discomforts, occasional harassments and physical danger that working on ships entailed. Maud earned far more as a shipboard masseuse than she could garner back in London, and she needed the money to support her illegitimate son, who was being brought up by his maternal grandmother. Violet's wages and hard-earned tips were keeping her ailing mother and younger siblings housed and fed. Economic necessity dictated their day-to-day decisions, so, despite their traumatic experiences, both women put aside any misgivings to resume their careers afloat.

Neither Violet nor Maud was asked to give evidence at the British or American enquiries into the *Titanic* disaster, which was perhaps a blessing, as it granted them some anonymity in later years. One prominent couple who did testify found their actions on the night publicly criticised, and their reputations irreparably tarnished. The clothing designer and successful businesswoman from London, Lady Lucy Duff-Gordon, had acquired first-class tickets for the *Titanic*'s maiden voyage, as she was travelling to New York to supervise her expanding fashion business. With her was her urbane and wealthy husband Sir Cosmo, and her maid, Miss Francatelli. The three of them were helped into Lifeboat 1, which had capacity for forty, but which was launched with only twelve people on board, mostly crew members. While they watched the *Titanic* sink, Lucy commented that her maid's beautiful nightgown had been lost. Understandably, this insensitive remark angered the sailors in the boat, who had lost both their shipmates and their livelihoods in the catastrophe. To calm them, Sir Cosmo offered to give them £5 each in compensation, so that they could buy new kit and clothes, but after they were rescued by the *Carpathia* and he was seen writing cheques to each of the sailors, malicious rumours circulated that the Duff-Gordons had bribed the occupants of Lifeboat 1 to ignore the cries for help from those drowning, in order to ensure their own survival. It was a complete fabrication, but the slur dogged the couple for the rest of their lives.

The *Titanic* was a colossal disaster, covered worldwide in the press. In the first week after the sinking, the *New York Times* alone dedicated seventy-five pages to coverage of the story. For those looking for a moral, the calamity smacked of hubris, a metaphor for the end of the expansionist, confident, entrepreneurial mood of the Edwardian era. Sir Osbert Sitwell, later recalling the changing atmosphere before the cataclysm of the Great War, saw the sinking of the ship as 'a symbol of the approaching fate of Western Civilisation'.[4]

Despite its brash emphasis on speed and its apparent reliance on technological wizardry, the *Titanic* fell victim to multiple human errors coupled with poor judgement. It flaunted all the trappings of opulent luxury for those who could afford it, but the sinking of the great ship on its maiden voyage across the Atlantic came to be seen as the moment harsh reality collided with the *belle époque* ideal of the Edwardian era. The brazen disparity in personal wealth exemplified by the ship also made uncomfortable reading for those with more egalitarian principles. It was reported that a de luxe suite on the *Titanic* cost $4,350, approximately $70,000 today. By contrast, the younger of the two Marconi men on board, Harold Bride, was paid a mere $20 a month. Without his determination to continue transmitting distress messages until the ship sank, the 706 souls eventually saved by the *Carpathia* might also have perished.

There was international public outcry about the injustice of how the 1,517 dead passengers and crew met their ends. Most of the 885 male crew members came from Southampton; 693 of them were lost, which was a devastating blow to a single city, and there was hardly a working-class home in the port unaffected by the tragedy. There were 201 survivors from the 324 people travelling in first class, and 118 of the 277 in second class, but only 181 survived from the 708 in third class. Although both official enquiries concluded that there had been no discrimination against third-class passengers, two witnesses testified that crew members had barred them from climbing to the upper decks where the lifeboats were housed during the crucial hours between the collision and the ship's sinking.

There were immediate repercussions for the travel industry, with a new emphasis on safety. Cunard reminded its captains that icebergs were a danger to all vessels on the North Atlantic in the spring, when increasing temperatures could cause ice floes to break away and float eastwards into shipping lanes. Better methods of safely launching lifeboats from listing ships were developed, including a design for an improved davit, the patent for which was filed by the author's great-great-uncle, Cunard Chief Officer Stephen Gronow. The number of lifeboats provided on the *Olympic*, *Titanic*'s twin, was increased overnight from twenty to sixty-eight. In future, lifeboat places were guaranteed for all on board, and practising lifeboat drill became an essential part of the voyage. All vessels carrying more than fifty passengers had to be equipped with long-range, permanently manned Marconi sets, which should be staffed round the clock so that vital messages and distress signals could not be overlooked. Watertight compartments were introduced throughout the hulls of existing and new ships, to limit the ingress of water and prevent the rapid flooding that had caused the *Titanic* to sink.

Following the demise of the White Star Line's flagship, all the rival shipping companies were keen to stress their vessels' safety and reliability, to woo back nervous passengers who had read the gruesome news coverage and the sobering findings of the official enquiries. It was vital to reassure the paying public – particularly female passengers, who were deemed to be naturally more trepidatious – and the lessons learned from the loss of the *Titanic* were incorporated into the design of Cunard's next ship, the *Aquitania*, launched in 1914.

The *Aquitania* was already under construction at the John Brown and Company works on Clydebank when the *Titanic* was lost in 1912. As a result, the structural design was substantially modified to include new safety features, and consequently the ship took more than three years to complete, being delivered in May 1914. The new ship had a double hull, and watertight compartments so that it couldn't sink as the result of a simple collision. It was the largest ship yet, half as big again as the *Mauretania*, and boasted four funnels and four direct turbines. The accommodation was 618 in first class, 614 in second and 1,998 in third class. The ship consumed 880 tons of coal per day, and had an average speed of twenty-three knots (about twenty-six miles per hour). It was fitted with Frahm's anti-rolling tanks to add stability to its performance in rolling seas, and it was the first Cunard ship to boast a permanent swimming pool.

This new Cunarder was the first vessel to become known as the Ship Beautiful. Her exterior was a triumph of marine engineering, while the 'backstage' elements of the ship, the places where only crew or officers would normally go, were the acme of the industrial aesthetic – hard-edged, functional, clean and compact. The *Aquitania* was particularly designed to be 'safe', and the company reiterated this word in its publicity material to attract and reassure passengers: 'The ship is fitted with all the latest safety appliances that facilitate safe navigation, and the great army of lifeboats includes two motor boats fitted with wireless … the Aquitania answers in supreme degree the requirements of safety, seaworthiness, luxury and comfort.'⁵

The public spaces and staterooms were deliberately designed to appeal to women. Arthur Davis, architect and designer, was briefed by Cunard that his role was to disguise from the first-class passengers – especially the female ones – the essential but unpalatable truth that they were on a huge vessel, afloat on an even vaster ocean, with all the attendant risks and discomforts. The greatest amenities and comforts to be found in the plushest hotels in America and Europe, such as the Waldorf-Astoria or Ritz-Carlton, were to be deployed on board, while the aesthetics should mimic reassuringly grand, long-standing, reliably *terrestrial* settings, in defiance of the maritime realities to be observed through any porthole. Davis recalled:

> I said to the directors of the company that employed me: 'Why don't you make a ship look like a ship?' ... But the answer I was given was that the people who use these ships are not pirates, they do not dance hornpipes; they are mostly seasick American ladies, and the one thing they want to forget when they are on the vessel is that they are on a ship at all. Most of them have got to travel and they object to it very much. In order to impress that point upon me, the company sent me across the Atlantic. The first day out I enjoyed the beautiful sea, but when we got well onto the Atlantic, there was one thing I craved for as never before, and that was a warm fire and a pink shade.
>
> The people who travel on these large ships are the people who live in hotels; they are not ships for sailors or yachtsmen or people who enjoy the sea. They are inhabited by all sorts of people, some of whom are very delicate

and stay in their cabin during the whole voyage; others, less delicate, stay in the smoking room all through the voyage … I suggest to you that the transatlantic liner is not merely a ship, she is a floating town with 3,000 passengers of all kinds, with all sorts of tastes, and those who enjoy being there are distinctly in the minority. If we could get ships to look inside like ships, and get people to enjoy the sea, it would be a very good thing; but all we can do, as things are, is to give them gigantic floating hotels.[6]

Cunard deliberately cultivated its American clientele, who represented 80 per cent of first-class passengers on its Europe-bound ships, and the *Aquitania* was labelled the Ladies Ship. Advertising stressed the convenience, safety and comfort to be found on board. Elite passengers were provided with a restaurant, grill room, drawing room, smoking room, lounge, a grand foyer, numerous salons and writing rooms, as well as the swimming pool and a gymnasium. There was even a garden lounge and a Historical Gallery of Art. Cunard was keen to stress a noble and venerable cultural history; the staterooms and first-class public spaces, such as restaurants and ballrooms, were designed to match the high expectations of seasoned hotel guests and regular international travellers. The liner was presented as the epitome of Ritzonia, a blend of the name of the hotel chain with the 'ia' ending that provided the distinctively sonorous names of each of Cunard's ships, all taken from ancient Roman provinces.

The second-class accommodation boasted three features usually available only to first-class passengers: a verandah café, gym and lounge. There was also a spacious dining saloon, drawing room and smoking room. The third-class quarters, while more utilitarian, provided large public rooms with specially designated open and covered decks and were a great improvement on the usual horrors of steerage.

In the public rooms and shared spaces the illusion was maintained that the passengers were guests staying in an aristocratic country house, one that reflected the spectrum of historical styles that might be found in a centuries-old mansion on terra firma. Anachronistic linenfold panelling and ornate marble fireplaces might seem incongruous in a massive steamship pounding across the ocean, but they gave a reassuring sense of solidity and reliability to the apprehensive traveller. Queen Anne-style libraries, Jacobean smoking rooms or Tudor grill restaurants, Georgian-inspired colour schemes, touches of *faux* Grinling Gibbons, with figured carpets and heavy velvet curtains recalled the familiar style to be found both in the Ritz and the private homes of millionaires. The air of opulence and tasteful entitlement was augmented by potted palms and ornate flower arrangements. These elements distracted the passengers from the truth, that they were travelling in a huge mechanical behemoth across thousands of miles of unknowable, volatile ocean.

The inaugural voyage of the *Aquitania* began at Liverpool on Saturday, 30 May 1914. There were only 1,055 passengers on board despite a total capacity of 3,200. Queasy superstition was now attached to the maiden voyage of an ocean liner. However, 2,649 had booked tickets for the return trip from New York. One of the passengers on the outbound journey, writing anonymously as 'W.A.M.G.', gave a jovial account of their departure in 'The Diary of an Aquitanian': 'Liverpool's crowd cheers and a flotilla of tugs and larger craft salute Britain's largest liner. We settle down to life on an ocean without waves. Much ado about cabins. The Cunard officials give a gyroscopic display of "suaviter in modo";

interpreted into the language of the Atlantic, this means every passenger gets the best cabin on the ship – result, excellent dinner eaten.' He commented on the stability of the ship, the outbreak of affability between the passengers, with new friendships being fostered over games of quoits, and he hints at romance developing between some of the younger passengers, with discreet dalliances on the boat deck after dark.

His account appeared in the ship's daily newspaper, the *Cunard Daily Bulletin*, which was distributed for the information and enjoyment of the passengers. It sold at two and a half pence, or five cents, and was compiled and printed on board overnight. The paper contained illustrated adverts for attractions such as Liberty's department store and high-end furriers in London, the Adelphi Hotel in Liverpool, and art dealers in major cities. The *Bulletin* provides a fascinating snapshot of the perceived interests of the moneyed passenger in the summer of 1914. One advert was particularly likely to appeal to any female traveller wary of thieves. The Keptonu Treasure Garter was a small chamois leather money belt, like a discreet purse, which could be held in place on the lower leg by a band of elastic, fitting just below the woman's knee. In an era of full-length skirts, this handy accessory was designed to keep a lady's valuables and jewellery safely out of reach of pickpockets, unless of course they were to start at one's ankles. Also featured in one issue was a transparently sycophantic photo showing two achingly fashionable young women at a racecourse, captioned 'Miss Nancy Cunard and Lady Diana Manners ... two of the prettiest and most popular of the younger generation in English society. Lady Diana is the youngest daughter of the Duke of Rutland and Miss Cunard is a descendant of the founder of the Cunard Line.' As might be expected, stories about travel featured strongly.

An interview with M. Louis Blériot, the pioneering French pilot, elicited his startling belief that transatlantic flight was almost within his grasp: 'Perhaps some day aeroplanes will be as commonplace as motor-cars are now,' he predicted. In another article, M. Blériot's great rival, Count Zeppelin, described his magnificent new airship, which was under construction in Germany, and stated that he would attempt the first transatlantic flight in it in 1915.

In addition, there were occasional daily news stories picked up and transcribed by the 'Marconi men', the on-board wireless operators. There were some nuggets of royal gossip, mostly focusing on possible dynastic intentions and the holiday plans of European royalty, many of whom were closely related to one another as descendants of Queen Victoria. On 22 June 1914 there was a report that King George V was planning to meet his cousin the Tsar of Russia in the autumn, renewing (completely unfounded) rumours that the Prince of Wales, the future Edward VIII, might be contemplating a marriage with one of the Romanov daughters. Two days later, Queen Mary's holiday plans were revealed: she was intending to spend part of the late summer of 1914 in Germany, staying with her aunt, the Grand Duchess Augusta of Mecklenburg-Strelitz, then at the summer residence of the King and Queen of Wurttemberg.

July 1914 marked a high point in transatlantic travel; trade was booming, with well-heeled passengers of all nationalities crossing the great gulf of ocean that separated the Old World from the New. The wealthy were able to enjoy conditions of unparalleled luxury and high-quality service and cuisine, all modelled on the best international hotels. The commercial shipping industry appeared to have learned valuable lessons from the hubris that caused the sinking of the *Titanic*, and the most modern ships, such as the *Aquitania*, were deliberately designed and marketed to inspire confidence in the most nervous passenger. Even those who had been personally involved in the tragedy, seasoned seafarers such as Violet Jessop and Maud Slocombe, had weighed up the alternatives, gritted their teeth and returned to shipboard life in order to support their dependants and maintain their households. On the upper decks of the great ships, stewardesses catered adeptly to the whims and needs of their ladies, alert that attentive service could be rewarded with lucrative tips. Meanwhile, many decks below, in steerage, experienced matrons kept an eye on the physical welfare of their polyglot emigrants, and prayed for a flat sea and a welcome breeze to improve the fuggy conditions in their tightly packed quarters.

No matter what their position in life, the hopes and aspirations of millions of people were about to be overturned by the accelerating pace of global events, which would lead to a great catastrophe. The first sign of the gathering storm can be seen in the *Cunard Daily Bulletin*. On 3 July 1914, on page 11, alongside entertaining accounts of American rowers at Henley and truculent suffragettes fighting with court officials in Caernarvon, the newspaper ran a brief news item that struck a discordant and sombre note. The bodies of the Archduke Franz Ferdinand, heir to the throne of Austria-Hungary, and his wife, both of whom had recently been assassinated at Sarajevo, were being transported from Trieste to Vienna. The Kaiser had decided not to attend their funerals, because of fears for his safety from anarchists. Within a month, the *Aquitania* had completed its third transatlantic trip, and Europe was plunged into war.

2

From the Ritz to the Armistice

———

Violet Jessop's life was immediately disrupted by the declaration of war on 4 August 1914. She was working as a stewardess aboard the White Star Line's *Olympic*, the sister ship of the *Titanic*, which was sailing to New York. The outbreak of hostilities surprised many merchant vessels already embarked on scheduled voyages, or berthed in what were, suddenly, hostile ports. Ocean liners hurried across open waters, fearful that enemy ships might already be lying in wait for them. On the transatlantic routes there was a great homeward rush of Americans and Europeans travelling in both directions.

Aboard the *Olympic*, Violet noted that the passengers managed to maintain the civilities but there was considerable tension among the multinational crew. The journey back to Britain in the huge ship, now darkened in case of enemy action, was sinister and menacing. Violet recalled: 'It seemed uncanny, journeying back to England, to goodness knows what, in a huge darkened *Olympic*. The first sign of a large ship with the unmistakable signs of a cruiser about her did make our hearts beat swifter. It was the Cunarder *Aquitania*, already converted to an armed merchant cruiser, looking very much like a woman showing off a beautiful new dress, which we duly admired, gratefully.'[1]

Passenger numbers were adversely affected by the outbreak of war, and many women seafarers were rapidly made redundant. Cunard's internal records show that by 1915, owing to the submarine menace, passenger traffic across the Atlantic had to a large extent been diverted to steamers of a neutral flag. However, there were still some Allied passenger ships plying the Atlantic, albeit less frequently, and it was generally believed that the Germans would not dare to target any passenger ship carrying civilians from a neutral country such as America or Canada. But those assumptions were proved catastrophically wrong by the sinking of the *Lusitania*, a huge Cunard liner, sailing from New York to Liverpool via Queenstown, in May 1915. On board were 1,959 people, including twenty-one stewardesses, a matron and a typist.

On the morning of the *Lusitania*'s departure from New York, on 1 May 1915, the German Embassy in Washington placed warnings in major American newspapers:

> Travellers intending to embark on the Atlantic voyage are reminded that a state of war exists between Germany and her allies and Great Britain and her allies; that the zone of war includes the waters adjacent to the British Isles; that, in accordance with formal notice given by the Imperial German Government, vessels flying the flag of Great Britain, or any of her allies, are liable to destruction in those waters, and travellers sailing in the war zone … do so at their own risk.

The threat of attack was not taken seriously by those booked to travel on the *Lusitania* because, unlike other Cunard vessels that had been converted to military and naval use, the *Lucy*, as she was affectionately known, was still supposedly engaged solely in conveying passengers, and had not been requisitioned by the Admiralty for war service.

Nevertheless, Captain W.T. Turner took every precaution, with bulkhead doors and portholes sealed and lookouts posted.

The *Lusitania* was a modern and well-designed ship, with modifications added since the loss of the *Titanic*; she carried forty-eight lifeboats, which were capable of accommodating 2,600 people, ample capacity for the full complement of 1,962 passengers and crew on board. However, on this voyage the ship was not travelling at full speed; due to the general decline in the volume of transatlantic travel and the increased price of coal, as well as a shortage of labour, it had been decided to reduce the steam power of the propellers by approximately a quarter, which simplified manning the stokeholds but reduced the speed of the vessel to twenty-one knots. Crucially, the *Lusitania* was moving slowly enough to provide a tempting target for the German U-20 U-boat, when she was spotted on 7 May 1915, ten miles off the Old Head of Kinsale in Eire.

The lookout reported an oncoming streak of foam and the ship was hit seconds later below the bridge by a torpedo. There was hardly any time to launch the lifeboats and evacuate, as the ship sank a mere eighteen minutes after the attack. Survival was random, a matter of luck. Although the captain insisted 'women and children first', as on the *Titanic*, so quickly did the *Lusitania* start to sink that panic spread and there was a mad scramble for the lifeboats, with the youngest and fittest inevitably winning the struggle.

Families were separated in the chaos; Lady Allan, wife of a partner in a shipping company, escaped in a lifeboat with both her maids and her Cartier diamond and pearl tiara, but her two teenage daughters, Anna and Gwendolyn, were among those who died because they were in a separate part of the ship when the torpedo hit and could not be located in time.

An Irish passenger in third class, Mrs Nettie Mitchell, found herself one of eleven people in Lifeboat 14, which was lowered successfully, but rapidly filled with water as it had no boat plug in place. The lifeboat was swamped, and sank almost immediately; her husband Walter and baby son were quickly drowned. Nettie slipped into unconsciousness, was picked up by a rescue boat and was left on the quayside at Cobh in the open air among a pile of corpses, as her rescuers assumed she was dead. By chance, Nettie's brother John, who had also survived the sinking, was searching the port looking for his family; he found Walter dead, then spotted Nettie's apparently lifeless body nearby. He noticed her eyelids were fluttering, and summoned medical help. Nettie was resuscitated, and recovered from her physical ordeal, though her double bereavement – the loss of both her husband and son in such traumatic circumstances – affected her mental health for some time. Eventually, her doctor recommended she take up some absorbing, worthwhile career as therapy; Nettie moved to Dublin and trained as a midwife, a vocation that she found greatly rewarding.

Of the twenty-one stewardesses on board the *Lusitania*, only eight survived. One of the fortunate few was Fannie Jane Morecroft, who had worked aboard the *Lusitania* since 1912. An ebullient Londoner who had a lively past and a spirited personality, she had eloped at the age of eighteen with a much older man, whom she married despite her parents' disapproval. In 1901 he died, leaving her a widow with two young children, and so she was forced to place her children with a foster family in Liverpool and to seek work at sea as a Cunard stewardess to support them. In later years she lied about her age, claiming to be seven years younger than she was in reality. Her great friend, Marian 'May' Bird, who was forty, worked alongside her on the ship, and they were both apprehensive of German submarines, which were known to be active around the coast of Ireland.

As soon as the torpedo struck, Fannie Jane Morecroft ran to the cabins in her care and hurriedly helped the women and children passengers into life jackets, then bustled them up on to the starboard deck. She graphically recalled that everyone was running about 'like a bunch of wild mice'. She assisted passengers into lifeboats, and watched as the boats were launched. Marian emerged from the crowd as the ship listed alarmingly and they began to slide across the tilting deck. The two stewardesses jumped overboard together and were pulled from the water by the occupants of Lifeboat 15, which was floating nearby. However, that boat became entangled in the exposed wires and trailing ropes of the stricken ship, and it was only by luck that at the last moment it was able to float free, rather than be dragged under as the *Lucy* sank. All those in Lifeboat 15 were picked up by a trawler and taken to Queenstown. So determined was Fannie to return to her family in Liverpool that, three days after the sinking, she arrived at her married daughter's home, still wearing the bedraggled stewardess's uniform in which she had been rescued.

Of 1,959 people on board, 1,198 were lost; there were only 761 survivors, and 885 bodies were never recovered. The sinking of the *Lusitania* caused international condemnation and outrage. British public opinion was incensed:

> The simple fact was that the *Lusitania*, carrying about two thousand passengers, who for purposes of interest or

pleasure pass to and fro across the Atlantic, was torpedoed without the usual warning, so that there was no time to rescue any but a comparatively small portion of those on board, and the hostile submarine made off without an attempt at rescue ... We can scarcely imagine even the bloodthirsty Hun gloating over the deaths of these.[2]

The day after the sinking, on 8 May 1915, Cunard's chairman A.A. Booth wrote a heartfelt letter to the company's general agent in America, Charles P. Sumner: 'We are all at one in our feelings with regard to this terrible disaster to the Lusitania, and it is quite hopeless to try to put anything in writing. My own personal loss is very great, as my New York partner, Mr Paul Crompton, with his wife and five children, all appear to have drowned ... the loss of life appears to have been appallingly great as the time was so short [between the torpedo's impact and ship's sinking just eighteen minutes later].'[3] The Board Minutes from Cunard on 20 May 1915 recorded the members' 'deep sense of horror at the outrage committed against this ship'.

The German government defended its actions against international denunciation, and, despite the findings of the commission appointed to investigate the matter, that there was no truth to allegations that it was a cruiser carrying troops and munitions, Admiral Tirpitz maintained in his memoirs that the *Lusitania* 'figured as an auxiliary cruiser in the British Naval list', and he described the ship as an 'armed cruiser, heavily laden with munitions'.[4]

The sinking of the *Lusitania* was the inspiration for one of the most compelling propaganda posters of the Great War. In June 1915 artist Fred Spear portrayed a drowned mother, her dead child still clutched in her arms, sinking down through murky waters, with the simple instruction 'Enlist'. It was an emotive and influential image, designed to inflame a sense of moral outrage at the atrocity perpetrated by the enemy, and contributed to America joining the Allies in 1917.

For women seafarers, as the Great War progressed there were diminishing opportunities to pursue their former careers afloat. So great was the public revulsion at the fate of the women and children killed by the sinking of the *Lusitania* that Cunard was reluctant to employ any female staff on ships until the end of the hostilities. Fannie Morecroft was laid off, to her great indignation, and she spent the rest of the war working in a variety of jobs, including that of tram conductor, though she was able to return to her maritime career when peace was restored. And she was not alone; as transatlantic passenger traffic shrank, many stewardesses of all lines were forced to seek employment in other fields. Some retrained as nurses, and found themselves working on one of the seventy-seven converted liners that were now hospital ships, where their sea knowledge was highly regarded. During the Great War, Queen Alexandra's Royal Naval Nursing Service alone had eighty-five sisters on thirteen hospital ships, and Dr Jo Stanley estimates that there were some 2,000 nurses employed on British hospital ships, coping with the wounded of all nationalities.

Nursing was to offer a worthwhile career to many a British woman who was previously denied a role outside the home. Sheila Macbeth Mitchell was born in Bolton to middle-class Scottish parents, and had hopes of becoming a teacher of physical education, but was dissuaded by her parents, who did not want her to work. The outbreak of the Great War swept aside such misgivings, and Sheila trained to become an auxiliary nurse in Queen Alexandra's Nursing Service. In 1916 she found herself on the hospital ship *Britannic*, working alongside veteran former stewardess Violet Jessop, and the two women left gripping eyewitness accounts of the disaster that befell the vessel. Violet's wartime transatlantic voyages had left her in no doubt of the value of nursing as a career. In the first months of the war, while she was working as a stewardes on the *Olympic*, commanded by the redoubtable Captain Haddock, their vessel was involved in the rescue of the crew of the Royal Navy battleship HMS *Audacious*, which had struck a German mine on 27 October 1914, between Northern Ireland and Scotland. Despite the appalling weather and mountainous seas, all the crew were saved, though the ship was lost. It was this experience that made Violet decide to train as a nurse, like many other young women. She joined the VAD, the Voluntary Aid Detachment. This was a nursing organisation that had been set up by the British Red Cross Society, with great success; by the end of the war there were more than 126,000 VADs working in support of the war effort, and they were greatly needed. Violet's four brothers were now serving in the trenches, and she lived in fear that one of her injured patients might turn out to be her sibling.

By 1916 Violet was a qualified nurse, but her considerable pre-war maritime experience as a stewardess made her doubly valuable on the HMHS *Britannic*, which had been despatched to the Mediterranean, to the Aegean Sea. There were 673 crew on board and 392 hospital nurses. The ship was enormous, a former White Star ocean liner designed for the transatlantic route, and it was the sister ship to the *Olympic* and the ill-fated *Titanic*, on both of which Violet had previously worked. The *Britannic* was charged with caring for wounded Allied servicemen, and was equipped with 3,000 hospital beds. Fortunately, it had not taken on any patients by the morning of 21 November 1916, when there was suddenly an explosion, caused by either an underwater mine or a U-boat torpedo, and the ship began to list.

Sheila related how those on board remained calm, picking up their life jackets and queuing quietly to enter their designated lifeboats. The order was given by the captain to evacuate, and the sixty-year-old chief matron, Mrs E.A. Dowse, stood calmly on the ship's deck, counting off each of the nurses as they filed into lifeboats with crew members. Mrs Dowse had been one of four nurses who had been in the 1885 Egyptian Campaign, serving in the Relief of Khartoum. During the Siege of Ladysmith in the Boer War, she was in charge of the hospital, apparently impervious to the constant danger from stray bullets.

The *Britannic*'s engines were still running and the ship was moving, as the captain was hoping to beach the vessel in the shallows. But the prow of the ship gradually filled with water and tilted downwards, while the giant propellers continued to turn, rising from below the sea to scythe through the air. As the first lifeboats full of evacuees struck out from the ship, relief at their escape turned to horror as boat after boat floated within reach of the vast, churning steel blades above them. Some seamen wrestled with the oars, desperately trying to steer their boats out of danger, while others leaped into the water, capsizing their boats, and were drawn inexorably into the bloody carnage left in the *Britannic*'s wake. Fractured wooden spars, life jackets, horribly injured men and women were all churned up in a bloody froth. Two lifeboats were cut in half, their occupants slashed to pieces in seconds.

'Unsinkable' Violet Jessop was the only survivor from her boat. Despite being unable to swim, she was wearing a cork life jacket, so she dived into the water as the boat overturned, and came up under it, sustaining a bad blow to the head. In the darkness her hand grabbed another's, and she was dragged to the surface by a surgeon from the Royal Army Medical Corps. There they floated among a sea of bloodied, injured and dead men. Violet had received a deep gash to one leg, a head injury and a fractured skull, but she was lucky to be alive. 'All the casualties were caused directly or indirectly by the propeller, and the wounds and fractures were terrible,'[5] recalled Sheila Macbeth Mitchell.

The survivors were picked up from the water, and the nurses in other boats tended to them by ripping up their aprons for bandages. The *Britannic* sank exactly fifty minutes after it was hit, and 1,035 survivors were eventually rescued by three British ships who, alerted by radio, had raced to the scene. The casualties would undoubtedly have been far higher had the ship been carrying sick and wounded soldiers when the incident occurred, as it would have been almost impossible to evacuate any bed-bound patients successfully. Thirty people died at the scene, but many more suffered life-changing wounds from the propellers.

While *Britannic*'s rescued male crew and officers were returned promptly to England, arriving in Southampton on 4 December, the nursing staff, including Sheila, were stranded for weeks on the island of Malta, unable to return to Britain. They were finally sent back on a former French cargo ship, an uncomfortable and seemingly interminable journey. They arrived at Southampton at last on Boxing Day and took the train to Waterloo Station, where their matron-in-chief ordered them to go home, rest and await further orders. Sheila was eventually posted to work in an army hospital in France, to her relief, as she was glad to be back on dry land. After a lengthy period of recovery in Malta, Violet was sent back to Britain via Sicily and through Italy by train. On her return she accepted that there were now almost no posts for women afloat while hostilities continued, so she put her linguistic knowledge to good use and took a job at the London branch of the Banco Español del Río de la Plata of Buenos Aires. It was as a result of this traumatic episode aboard the *Britannic*, her third maritime catastrophe, that Violet gained the soubriquet of The Unsinkable Stewardess. She wore a wig for the rest of her life to conceal the injuries sustained to her scalp.

The *Britannic* was the largest vessel to be sunk during the First World War, and remains the largest passenger liner ever sunk. In a curious footnote, in 1976, when she was eighty-six, Sheila Macbeth Mitchell responded to a worldwide appeal for witnesses to the sinking of the *Britannic* from French naval officer Jacques Cousteau, who wanted to explore the recently discovered wreck on the seabed. Sheila not only supplied him with uniquely detailed information about how the vessel came to grief, but she also flew to the Aegean and descended to the seabed in a miniature submarine with the divers, though she feigned disappointment that they were unable to retrieve the alarm clock she had left on the bedside cabinet of her cabin sixty years before.

Given that many of the ocean-going ships were deemed male-only environments while hostilities lasted, in November 1917 the Royal Navy created the Women's Royal Navy Service (WRNS) – a body later known as the Wrens – to provide land-based support for the men fighting at sea. The female workforce ashore was also galvanised to support the war effort, particularly in support of the beleaguered merchant navy

By the end of August 1914, 30,000 men were volunteering to join the armed forces every day, and there was a desperate shortage of skilled labour as they left their previous occupations for military training. In keeping with the national drive to free up the male workforce for the armed services, women were recruited for manual jobs, including working in the shipyards, which was a radical departure from the pre-war expectations of genteel paid employment deemed suitable for the fairer sex. A newspaper article on women shipbuilders grittily reported:

> They are working in blacksmiths' forges; they red-lead iron work, and do certain portions of the paint work. All over a shipyard they may be seen tidying up, shifting scrap iron, carrying baulks of timber, pieces of angle iron, and iron bars ... A more valuable part of their work perhaps is done with machinery, especially in the joiners' shops ... In the engineers' section of the shipyard, also, they work screwing and boring machines, sharpen tools, and in many other ways help in this department. Experienced girls are very skilful in manipulating such powerful machines as those used for cutting angle-iron and for keel-bending. They drive electric cranes and winches, work which demands the greatest steadiness and care. The proportion of women employed in engineering works is greater than that in the shipyards, on account of the larger number of machines available. As time goes on, and the value of women to the shipbuilder grows, the percentage

of women workers engaged in the construction of merchant steamers will doubtless increase.[6]

Following the outbreak of war, Cunard continued to service and repair its own ships, but also supported the war effort by servicing and maintaining naval craft in its dockside facilities. Cunard already ran a complex, multi-faceted shore-based operation; it owned extensive premises on land, warehouses, repair shops and engine works, and employed hundreds of highly efficient managers, clerks and administrators, all based ashore but working in support of its ocean-going fleet. In October 1915 the company converted its Branch Engine Shop in Bootle into the Cunard National Shell Factory. Managed by Alexander Galbraith, who had the pioneering idea of recruiting some 900 local women, as well as 100 men to work there, the factory produced the first 6- and 8-inch shells made in Britain using female labour, along with a variety of other weaponry. As a bold experiment in bringing women into munitions manufacture, the Cunard Shell Factory was a triumph, and it was much imitated; by 1918 there were nearly 1 million women employed in British munitions and engineering works. The Lady Shell Workers, as they were known, were committed to supporting the war effort in their limited free time, putting on concert parties for wounded soldiers and performing at fundraising events, at which a spirited rendition of 'God Save the King' always closed the concerts.

Meanwhile Cunard was given the task of establishing and running the National Aircraft Factory Number 3, to build planes for the Royal Flying Corps and the Royal Naval Air Service, which amalgamated in 1918 to become the Royal Air Force. Forty-one per cent of the 2,600-strong workforce were female; photographs of the 'covering room' show scores of women dressed in pinafores and caps, busy attaching canvas to the wooden frames that formed the wings of planes. The factory constructed 126 fighter planes by 1919, a large contribution to the fleet of the new RAF.

On Armistice Day, 11 November 1918, at 11 a.m., the guns finally fell silent, marking the end of fifty-two months of carnage and slaughter. The Shell Factory closed days after, having manufactured 410,302 missiles. Three weeks later, on the evening of 7 December, a grand farewell party was held for the women employees. There were speeches of thanks and celebrations, the guests danced to the Cunard Orchestra, and it was announced that King George V was to be presented with a book of documentary photographs taken at the Shell Factory, at his request. The Lady Shell Workers received letters of heartfelt thanks for all their war work from Sir Alfred Booth, the Cunard chairman, and 'a memento of your share in the great victory ... a souvenir in the form of a 4.5″ H.E. shell, with the compliments of the Cunard Steam Ship Company'.

For Violet Jessop, now working for a bank in the City of London, the formal end of the hostilities was a hollow triumph. One of her beloved brothers had been killed in the closing days of the war, and she did not have the heart to join the seething and excited crowds out in the street. She recalled:

> I went into the manager's office with a letter I had been translating for him. We were the only two left in the building. There were tears in his eyes. We wept openly as we discussed the contents of the letter. His old heart was torn by the news that his youngest son had just been killed

in action during the same engagement as my poor Philip. We had our grief in common, and to neither of us did that Armistice Day bring a message of joy.[7]

More than 9 million men had been killed in the Great War, 942,135 of them from the British Empire. During the four years of conflict, Cunard had lost twenty ships through enemy action, including the *Lusitania*, the most high-profile passenger ship to be sunk during the First World War. The final tally represented 56 per cent of the company's pre-war tonnage. Even the *Carpathia*, which had steamed to the rescue of so many *Titanic* survivors in 1912, was sunk. Cargo vessels bringing vital supplies to Fortress Britain were also decimated by the German U-boat fleet. Among Cunard's many losses was the *Vinovia*, sailing from New York laden with brass and munitions, which was torpedoed and sunk in the English Channel on 19 December 1917, with nine crew members killed. That vessel was commanded by Captain Stephen Gronow, the author's great-great-uncle; he survived a night in the open sea, and was rescued the following morning, unconscious and suffering from hypothermia, but he recovered and returned to service on the *Aquitania* when peace returned.

While the material losses were considerable, it was the human losses that were the hardest to bear. Some 650 of Cunard's crew members and officers were lost at sea in the hostilities. A source of pride, however, was the way in which the company had coped. The ships had covered more than 3.5 million miles as part of the war effort, carrying 9 million tons of food, munitions and raw materials. Their main area of operation was the North Atlantic, but they also travelled to the Mediterranean and the far north of Russia. Cunard alone transported 900,000 troops, half a million of them American soldiers joining the war in Europe.

But the victorious nations felt little in the way of triumph, just exhausted relief, as they attempted to make some sense of the peace. In Britain there was hardly a household that hadn't suffered a bereavement or serious injury. There was a palpable absence of old pals and contemporaries, 'missing in action' or 'lost in France'. The survivors of the four years of war were now returning, some of them maimed, gassed, blinded or disfigured, or mentally affected by their experiences.

With hundreds of thousands of soldiers and sailors needing repatriation, the ocean liners that had served as troopships during the war were once again pressed into service. Surviving ships were refitted and once again put on to passenger routes, as shipping companies commissioned new vessels to replace their lost fleets. With so many people on the move between continents, it was inevitable that they would be taking with them more than memories and souvenirs; the Spanish flu pandemic, which swept the world in 1918–19, killed an estimated 50 million people worldwide, approximately 250,000 of them in Britain.

For women the first year of peace was a period of transition. Injured or traumatised ex-servicemen returned to Britain after the war expecting to take up their old jobs; indeed, many of them had been promised by their employers that they could return to their previous posts after the victory. While the men had been fighting in the trenches, many Civvy Street roles had been largely occupied by women. It was a thorny problem: the women who had been recruited and welcomed into office jobs in order to 'free up a man for the front' were now surplus to requirements. Cunard's staff magazine, *Cunard Line*, lost no time in dropping unsubtle hints to its female workforce: 'The day is coming soon when the rightful inhabitants will be welcomed back to the office and many of the ladies must lay down their pens sadly – yet gladly – and retire into private life. May it truthfully be said on that day that they have carried on.'[8]

While individual women might be prepared to relinquish a specific 'man's job' to a former soldier returning from the front, in general they had no inclination to return compliantly to the domestic hearth after the Great War. A month after the Armistice, in December 1918, British women were granted the vote for the first time (though not all women: for the next decade, enfranchised females were only those who were householders and aged thirty or over). This was both a concession to the pre-war impetus to introduce women's suffrage, and an acknowledgement of the vital role women had played in support of the war effort. The fundamental changes to women's lives were more profound than gaining the vote. Four years and three months of hostilities had transformed the nature of British civilian society. For the first time, large numbers of women of all classes and backgrounds had found useful, interesting and remunerative employment outside the home. For younger women, the Great War had overturned all their previous assumptions, and revolutionised every notion of how their future lives might be lived. Galvanised by the war effort and encouraged to demonstrate their patriotism, they had learned to drive motor vehicles, helped to construct munitions and aeroplanes, and trained as nurses or stenographers. For the first time women had been actively recruited and welcomed into offices and banks, factories and canteens. They had been employed alongside men, in clerking and administrative positions. Their abilities and skills had been required and valued, because they were filling vacancies left by men called up for the forces. As a result, many young women had developed a new sense of independence and self-confidence. They had a sense of their own agency, and had proved to themselves that they could earn an independent living. The most outgoing and enterprising were realising that they could be citizens of the world, especially if they were prepared to travel. And in order to do that, they had to go to sea.

3

Sail Away: Post-war Migration and the Escape from Poverty

In the immediate aftermath of the Great War there was a considerable demand for all forms of travel, and people on both sides of the Atlantic were desperate to take to the seas again now that the only perils they faced would be natural ones – storms, mountainous waves, icebergs and seasickness – not enemy torpedoes or floating mines. Naturally, there had been passengers who for a variety of reasons had braved the Atlantic on Allied vessels, despite the risks from lurking U-boats, but the numbers had slowed to a trickle, particularly after the sinking of the *Lusitania*. There had also been wartime limits on movement for British civilians under the Defence of the Realm Act, combined with the strictures of rationing, but now would-be passengers could venture overseas once again. Many felt the need to escape the all-pervasive sense of gloom that seemed to hang over the British Isles like a pall, and those who could afford a ticket grabbed the first opportunity to book a transatlantic trip on one of the recently demobbed, hastily converted or newly requisitioned great liners.

To meet the surge in demand, the shipping companies hurried to locate and sign up their pre-war female workforce, and to replace those who had left the industry to pursue other opportunities when they had been laid off. Experienced stewardesses were particularly highly valued, especially if they had gained nursing qualifications in the interim, as many had. Perhaps surprisingly, considering her harrowing experiences, Violet Jessop gave up nursing and chose to return to the seaborne life as a stewardess in 1920, when the opportunity presented itself, and she was not alone. Fannie Jane Morecroft, survivor of the sinking of the *Lusitania*, resumed her career with Cunard in 1919, and went on to become the chief stewardess of the *Lancastria*, a position that brought her the considerable honour of a stateroom to herself and a further decade of lucrative employment at sea.

Violet and Fannie, like most female employees, had been made redundant by the shipping companies as the danger to Allied shipping increased, but a very small number of seafaring women had continued to sail throughout the hostilities, and they had accumulated war service stories of their own. Cunard publicly celebrated those who had been employed on its ships throughout the war years. One was a veteran stewardess called Miss J.S. Cole, who was torpedoed three times during the Great War. A profile of her appeared in *Cunard Line* in May 1921:

> Miss J.S. Cole, stewardess, RMS Caronia, whose photograph is reproduced on this page, had the unpleasant experience of being torpedoed three times during the hostilities. She was in the Alaunia when the ship, having landed her passengers and mails at Falmouth, after a voyage from New York, was torpedoed on her way to London, near the Royal Sovereign Lightship, on 4th October 1916. Miss Cole escaped the further attentions of the enemy until 4th February 1918 when she was one of the complement of the Aurania, which was torpedoed off the north coast of Ireland. In the following July

Miss Cole had her third baptism of fire, being stewardess in the Carpathia, which fell victim to a U-boat some 120 miles west of the Fastnet.

Two stewardesses on the *Saxonia*, Mrs Agnes Stevens and Mrs Emily Dawkins, were similarly featured in *Cunard Line*. Agnes was originally from the Isle of Man, had two sons and one grandchild, and had worked as a stewardess for eighteen years. She had been born on a ship owned by her father, and had spent most of her life afloat. Emily had seven sons and six grandchildren and had been a stewardess for sixteen years. Although it wasn't stated directly, as both women were styled 'Mrs', it was likely that both were widows left with children, hence their need to work on ships. They had served together as nurses at sea during the Great War, and proudly wore service ribbons on their uniforms. The inseparable cabin-mates recalled being torpedoed while working on the *Ausonia*, off Queenstown in Ireland, close to the site where the *Lusitania* was sunk:

> We were down below in our room, knitting. We heard the alarm, we climbed up and looked through the port, and we saw the thing – the torpedo – coming towards us. It struck us but we discharged a depth bomb and got the submarine. Of course, we passed through the submarine zone time after time, but we were never on a ship that sank. We used to tell the soldiers not to worry, that we were mascots, and that as long as we were on a ship it wouldn't go down [. . .]
>
> We were on the Saxonia when she took over the first detachment of American troops and we came back with her at Christmas time, just after the Armistice, when she brought back the first load of wounded Americans. We

have travelled when we were the only women on a ship full of soldiers. We punched their meal tickets three times a day, we mended their stockings, we served in the canteen.

The stewardesses were keen to stress that they enjoyed their jobs. 'There's the air and there's the water, and you don't have to worry about the high cost of living as you're afloat,' said Agnes, though she admitted she still suffered occasionally from sea-sickness.[1]

Newly recruited stewardesses were needed to cater for the booming numbers of female travellers from all backgrounds. On many west-bound transatlantic crossings, women outnumbered men two to one in all three categories of accommodation. There were complex social and economic reasons why European women of all classes were now taking to the seas, but fundamentally it was due to their need for economic security, coupled with a biological imperative to strike out in a bigger gene pool.

During the Great War many women who would never previously have worked outside the home had joined the labour force for the first time, a course of action unimaginable to them and their families just a few years before. Their patriotism had been stirred by the war effort, but they also enjoyed earning an independent salary, and the sense of camaraderie and social status that came with having a shared sense of purpose. However, with the return of peace, some harsh economic realities were unavoidable. Women were pressured to retire from the workforce as tens of thousands of demobbed ex-servicemen returned to Civvy Street, expecting to pick up their old jobs. Cunard's staff magazine revised its previous noble sentiments, and stepped up the pressure: 'The end of the month is not the great joy to all, as some would have us believe. To a number of the ladies lately it has meant "the parting of the ways". The ladies' staff is gradually decreasing, owing to the return of the men from H M forces, and those who so gladly came forward to help "carry on" are now quietly slipping back into their old places.'[2]

Returning meekly to one's 'old place' might suit conservative values, but it was incompatible with the practical problem of 'surplus women'. A million men had been lost in the conflict, and a generation of young British women now faced a future that might not include marriage and the secure financial future a husband would have provided. Those whose boyfriends, brothers, fathers, fiancés or spouses had died in the great conflict needed to find a way to lead financially independent lives. There were also war widows, often left with children, who urgently had to find some sort of gainful employment to support their families. Women wanted dependable careers that brought in sufficient funds to keep a modest household solvent, not just a hobby that provided pin-money. Cunard cannily advised its own 'bachelor girls' to learn a marketable skill they could take overseas:

> England will always be a country where there are more women than men, and the death toll taken of our men during the Great War will be felt for years to come. It is the duty at the present time of every English girl to give of her best talents and energies to her country, and not be content to work just from week to week for sufficient salary to pay for her keep, clothes and amusements.
>
> ... At the present moment, a capable woman, armed with shorthand, typewriting and a language, can travel the world, passage paid, and return to a comfortable post

in England when advancing age and a wish for comfort and affection, rather than constant change and interesting work, make home life more attractive.

In 1912 it would have appeared a very strange happening for a girl to take a post as a stenographer in Germany, France, South America, or to travel as secretary to a novelist, shipping magnate or film producer. In doing any of those things, a girl laid herself open to be considered fast or queer and her people were often blamed for allowing it. Now it is taken as a matter of course, and the restless, roving spirit some women possess is finding a very happy outlet in seeing new countries and faces.[3]

'New countries and faces' certainly appealed to many British women, especially those from the working classes targeted by recruitment agencies overseas, such as the Society for the Overseas Settlement of British Women. SOSBW catered for many unattached women and girls interested in trying their luck elsewhere within the British Empire.

In the early 1920s, 24 per cent of the total land mass on earth was part of the British Empire, and swathes of the globe were coloured pink, indicating the dominions, colonies and overseas territories administered by the United Kingdom. British expatriate families working in far-flung places required staff, and British-born nannies, governesses, housekeepers, chambermaids and cooks could earn three times their previous salaries working in Australia, New Zealand or South Africa. In addition, single women 'from home' were scarce in these communities, so there was a decent chance of marrying a bachelor or widower. A respectable colonial marriage for a former domestic servant offered the gratifying prospect of a comfortable life in a household, complete with her own staff.

The Servants Crisis of the early 1920s arose because British women were reluctant to return to their pre-war roles, carrying out dull, poorly paid drudgery in richer people's households. Domestic service could not compete with their more interesting and rewarding war work and any deference felt by the working classes for their social 'betters' was declining, as realisation grew that the upper classes' pre-war lives of leisure and privilege had been enabled and facilitated by their staff. The government offered meagre incentives to tempt women back into domestic service, such as free uniforms, but most refused to sign up, calling instead for proper vocational training for women in alternative careers, such as nursing, midwifery, comptometer training, or shorthand and typing. Many women also yearned for travel and romance, and dreamed of being 'spotted' by a talent scout or a movie director. Seeing such fantasies in the cinema, popular newspapers and magazines sparked a certain amount of restlessness. Those young women who had the choice would much rather be secretaries or typists, hairdressers or factory hands than scullery maids.

Genuinely wealthy employers, of course, could still attract servants by paying excellent wages, typically four times what they had paid in 1914, and also guaranteeing better conditions for their employees. But well-heeled households in America and Canada were also crying out for English-speaking staff, and there was considerable cachet in those societies in employing a British-born butler or a Scottish nanny. Recruitment agencies and resettlement organisations with a welfare agenda helped women relocate overseas, by finding them posts in foreign cities and assisting their ticket purchases. Thousands of British and Irish economic migrants, former shop assistants, typists, factory hands and domestic workers of all types, embarked on transatlantic liners, knowing that this really would be 'the trip of a lifetime'.

In the early 1920s the more fortunate women passengers heading west on the ocean liners had resolved to emigrate to North America, and they had some reasonable expectations that their life chances would improve by relocating, because they had been actively recruited by agencies and governments, and reassured by shipping companies and stories in the media. However, there were also tens of thousands of people from all over Europe and beyond who were heading for America under duress. Whole populations had been displaced by the First World War, communities had been uprooted by the conflicts that had swept across their nations, and families sundered by world events beyond their control. Great numbers of the determined and the desperate, filling the lowest berths in the ships, travelling frugally in third class, were all heading for new lives. Some were buoyed by optimism, while others were running away.

For many German women, travelling to America was a way to escape dire poverty. In the wake of the Great War, hyperinflation decimated the country's economy; the mark had stabilised at about 320 to the US dollar in the first half of 1922, but by the end of the same year it had sunk to 8,000 to the dollar. On 13 December 1922 the *New York Times* published a remarkable story about an intrepid German stowaway. Christiana Wilhelmina Ida Klingemann was a forty-one-year-old former stewardess who had spent eight years working on the Hamburg-Amerika Line. Unemployed since the Armistice, she had found it difficult to earn enough to survive, and she could not afford to feed herself, her invalid widowed mother and her brother. She feared they would all starve. A glimmer of hope was offered by one of her three cousins who were already settled in California; he was a teacher, living on a modest wage, and he promised that if she could get to New York, he could send her the train fare to the west coast. But it was a desperate gamble; unable to afford the price of even the cheapest third-class ticket to cross the Atlantic, Christiana felt she had no choice but to risk her life and stow away on a passenger ship bound for America. Her first attempt was on the *Wuerttemberg*, a ship run by the Hamburg-Amerika Line, her former employer. Although she knew the layout of the ship well, she was discovered by the crew and ignominiously returned to shore. On her second attempt, she took no chances: in order to evade discovery, she was prepared to conceal herself in the bowels of the ship, beneath the ballast in the hold.

Familiar as she was with the customary procedures and practices before embarkation, Christiana managed to slip aboard the *Pittsburgh*, a White Star Line steamship, and secrete herself in its hold, way below the waterline in the lowest level of the vessel. Two hundred tons of gravel had been shovelled into the hold to act as ballast to stabilise the ship, and Christiana lay down in it, covering her body with a layer of the small stones to avoid detection before the ship sailed. Following a cursory inspection, the hatches to the hold were sealed, and the ship left Bremen for New York. Christiana, in total darkness and completely alone, survived for days on black bread and sausages, hoping that the hold was free of rats. It was a dangerous gamble, and one that could have been fatal, as no one knew she was there. In addition, if there had been a winter storm – a real danger in December – she could have been badly injured by the turbulence of the waves, thrown against the bulkheads and pounded by the ballast. She had brought along a large bottle of water, but once that was gone and she felt sure the ship was more than halfway to its destination, she hammered on the inside of the hatch cover. Fortunately, a steerage passenger heard her above the noise of the engines, and he raised the alarm. The chief officer was summoned, and Christiana was retrieved from the hold after nearly a week, hungry and filthy, but able to tell her story coherently.

Christiana had brought a man's suit in which she had hoped to make her escape once the ship docked in New York. Although she had no passport when discovered, she had identification documents, including her service book showing her former employment as a stewardess. Her rescuers this time were more sympathetic, and she was supported in petitioning the Commissioner of Immigration, Robert E. Tod, to permit her to land at Ellis Island so that she could find work and send money to her desperate family.

The emigration experiences of Marie Riffelmacher from Altenburg in Germany were more orthodox, but similarly life-changing. Marie was fifteen in the summer of 1923 when she set out with her two older brothers, Matthias and Friedrich, to emigrate to the United States. They were sent to stay with cousins already living in Michigan, to try to earn enough money to enable their parents and three younger siblings to emigrate from Germany later. The tough decision to send them abroad was made because food was very scarce; their father, Leonard, brought home less than 2 million marks a week, and in 1923 a pound of meat cost 180,000 marks, a loaf of bread was 90,000 marks, a litre of beer was 30,000 marks and a single egg cost 15,000 marks. Various photographs of the Riffelmachers at home reveal how their fortunes had visibly declined because of the war: from a comfortable, bourgeois existence, they were now lean and hungry. Marie's

73

mother in particular looked gaunt and emaciated, with hollow cheeks; she had probably denied herself food to keep the children from hunger. Desperate measures were called for, and the Riffelmachers managed to raise sufficient money to buy three one-way tickets on the Cunard ship *Tyrrhenia*, which offered basic but congenial third-class accommodation, to send three of their children to the New World.

In an account written up by her granddaughter and deposited in the Cunard Archives in Liverpool, Marie recalled that the villagers of Altenburg gave them a resounding send-off, playing music and hymns. The teenagers travelled across country, then caught a train at Nuremberg to Hamburg, where they boarded the ship. It was a ten-day voyage to America, and they slept in bunk beds, with men segregated from women and children. There was ample food, which the Riffelmacher siblings enjoyed. Many fellow travellers suffered from seasickness, but Marie and her brothers had received a small jar of honey from a neighbour, who assured them it would cure seasickness, and perhaps it acted as a placebo, because they were completely untroubled by nausea.

On arrival at Ellis Island in June, fifteen-year-old Marie was intrigued by the 'nice lady', the colossal bronze figure of the Statue of Liberty, unaware of its significance. In her account she describes being examined and passed for health problems. She added that the Riffelmacher teenagers had no difficulties over money or luggage, as they didn't have any. Two things struck Marie as strange as she and her siblings waited to be processed by the immigration authorities: bananas, which she was given on arrival and which were for her a novelty; and people of different racial backgrounds, who she was seeing for the first time in America. The formalities completed, the family were allowed to travel to Grand

Central Station where they boarded a train bound for Bay City, Michigan. They arrived there at 4 a.m. on a Sunday, were reunited with their cousins, and by nine thirty that same morning they were in church.

Marie initially worked on her cousins' farm, then as a nanny for another relation. Despite the kindness of her relatives and her much improved circumstances, she was homesick and remembered receiving a letter from an aunt in Germany describing how much Marie's mother had cried after the siblings had left. However, the Riffelmacher family were not separated for long – in October 1923 Marie's parents followed her to the States, sailing on the Cunard ship *Laconia II*. A surviving photograph shows a merry-looking group of fifteen adults, some clutching musical instruments, and a number of children, posing for an unknown photographer on the small open-air third-class deck. Marie's youngest brother, Georg (aged three), is a small blond boy with a serious expression, posed in front of the ship's lifebelt. Unknown to the adults, little Georg was already gravely ill; he had been bitten by a dog in the street just before they boarded the ship, and he died of rabies just three weeks after the family arrived in the United States.

Despite the cruel loss of their youngest child, the Riffelmacher family settled in the States and began to prosper in a manner that would have been impossible in Germany. Marie worked as a maid for a family in Bay City. Conditions were good; she was paid $10 a week, and allowed two half-days' holiday a week. Occasionally she was able to send dollars back to her pastor in Altenburg; foreign currency was a great help to Germans still suffering hyperinflation. By the end of 1923, a single American dollar could be exchanged for 4.2 trillion marks. Marie married Harold Stroemer, and they had three children, nine grandchildren and eighteen great-grandchildren. She lived in Bay City, Michigan, all her life.[4]

In the years just after the war America offered a lifeline for hard-pressed economic migrants like Christiana Klingemann and Marie Riffelmacher. However, American attitudes were hardening against the previous liberal policy of allowing in people from the more remote and hard-pressed parts of the world. In the first years of the century, between 750,000 and 1 million Europeans a year arrived in America seeking citizenship. By the early 1920s many Americans were clamouring for restrictions on immigration because of rising unemployment; by 1921 there were more than 5 million Americans out of work. In response, Congress passed two immigration restriction acts, drastically cutting the numbers allowed into the country.

The US Immigration Act of 1924 was even more restrictive than that of 1921. The total number of European immigrants now allowed in annually was slashed to 161,500 in any one year. By comparison, in 1913, 1.141 million European immigrants had been admitted to the US. In addition, visas were now required, and these were issued by US Consular offices in the countries of origin. Any shipping line bringing in an immigrant without a visa, or surplus to the annual quota allowed to that country, could face a punitive $1,000 penalty. Immigration figures plummeted, though some nations were still favoured over others: British- and Irish-born immigrants were allowed a generous proportion of the annual European allocation, and anglophones with valuable skills, such as British-born servants or nurses, were actively encouraged to emigrate to the North American continent.

While restrictions were enforced to reduce immigrants'

numbers, paradoxically the authorities dealing with the thousands of people passing through their hands daily were treating them in a more sympathetic way. Frederick A. Wallis, the Commissioner of Immigration for the State of New York, introduced a more enlightened and humane regime on Ellis Island. Mandatory health tests were conducted on all incomers, and up to 85 per cent of them required detailed medical assessments before being allowed to enter the USA. Wallis was struck by many migrants' profound poverty; one woman, travelling with five barely clothed children, was planning to take them to Chicago. She had a dollar and eight cents, but no railroad tickets. Some steamships were notoriously filthy, so their passengers arrived soiled. Other lines were so rapacious that steerage passengers were only supplied with drinking water during the voyage if they paid extra for it. Wallis felt there was no excuse for the insanitary conditions, especially considering the exorbitant prices some companies were charging these desperate people. He also wrote movingly about the process of winnowing out those arrivals deemed too ill, old or feeble to be admitted under the tough new immigration criteria:

> Every day is Judgement Day for many people at Ellis Island … families are cut in twain, husband and wife separated, children taken from their parents, or one taken and the other left. It is all wrong … These people have been saving for years, denying their families many little luxuries in order that they might get together sufficient funds to come steerage. After years of sacrifice and saving, they come to this port only to be sent back to Europe. And sent back to what? Literally to the Devil and his angels.

Europe is worse off today than during the war. These people go back with no home, no business, broken in pocket, and, a thousand times worse, broken in spirit. No one can ever picture the scenes of anguish of spirit we see at this port. We frequently found it necessary to carry people bodily from the building and put them on the ship, many of them going into hysterics and threatening to jump overboard.[5]

Would-be immigrants could be deported if they showed signs of physical or mental illness, but less well-known is the discrimination against the illiterate. Adults arriving at Ellis Island were required to read out loud forty words of a printed language – any language; a wide variety of texts were supplied – to prove they were literate. This discriminatory policy often split up families, and it particularly affected women from remote and traditional communities in eastern Europe, who were less likely to have been taught to read. Wallis recalled one case where a young Jewish woman was parted from her younger brothers and sisters, and was to be deported because she couldn't read, though they could. Sobbing, she explained that she was the eldest and had never learned to read because she had to stay at home and work so that her younger siblings could be educated. Despite sympathy from the immigration officials, she was sent back to her home country and an uncertain future.

In the early 1920s poor immigrants hoping to gain perma-nent entry to America found the process increasingly difficult as quota restrictions and stricter entry criteria were enforced on those 'huddled masses, yearning to be free'. Paradoxically, for the more privileged women – crossing the Atlantic in the first- and second-class decks far above steerage, in every sense – their voyages between Europe and North America had never been so fascinating or enjoyable.

On the upper decks of every ocean liner, accommodated in luxury, were the leisured wealthy, who were destined to become the new 'cruising class' as the economic realities of ocean travel changed. The respectably prosperous who travelled in second class for business or commerce hoped to pick up the threads of their disrupted mercantile life or to create new fortunes in the New World. On the larger ships there was usually a 'floating population' of travellers hoping to profit from their more naïve shipmates: they were the serious gamblers, gigolos, ocean vamps, card-sharps, procurers, snake-oil salesmen and blackmailers. And then there were the performers – those already established as actors, singers or dancers – hoping to make it big on the stages of New York or London, or even on the silver screen, as well as aspirant stars whose names were only known to their nearest and dearest. The transatlantic ship not only provided a form of transport across the globe, it was also both the practical means and the symbol of opportunity, of new beginnings and fresh starts. From 'third class' to 'top deck', from desperate women escaping financial hardship to wealthy international sophisticates hoping for romance and adventure, every transatlantic liner of this post-war era was freighted with hope.

4

The Roaring Twenties

In January 1924 Parisian-born impresario André Charlot and twenty exuberant and sociable chorus girls and actresses boarded the *Aquitania* at Southampton in a high state of excitement. Following a successful London run, they had been booked to take *Charlot's Musical Revue* to New York, with an all-British cast. The *Revue* featured comic songs by Noël Coward and Ivor Novello; it relied upon ribald Cockney slang, and was peppered with London colloquialisms, but, contrary to most expectations, the show captivated American audiences. The company's initial six-week booking on Broadway was extended for a nine-month run, eventually totalling 298 performances. Gertrude Lawrence and Beatrice Lillie were in the *Revue*, and the success of the show subsequently brought them international fame as actresses.

As might be expected, the voyage out on the *Aquitania* was a particularly lively and entertaining one for the performers and their fellow passengers. The extrovert company were all excellent dancers, and every evening they took to the floor to dance with their fellow passengers, accompanied by the scintillating sounds of the ship's band, who probably couldn't believe their luck. Anticipating potential difficulties, Mr Charlot had wisely engaged a female chaperone to look after the young women's welfare and to curb their enthusiasm while on board, but despite Matron's best efforts, one ingénue managed to conduct an illicit romance with a millionaire, and accepted his offer of marriage, so their numbers were reduced to nineteen by the time they arrived in New York.

The liveliness of all parties on board the *Aquitania* during this voyage might be put down to natural causes, but it may also be attributed to high spirits, particularly alcoholic ones. In the 1920s European shipping lines had one considerable advantage over their American rivals: they were able to serve alcohol on board their liners. Prohibition – the banning of the import, manufacture and sale of alcohol – had come into force throughout the United States in January 1920, and it was to have a profound effect on many aspects of American life for the next decade.

The new law did not ban the actual consumption of alcohol, so those Americans with foresight and funds simply stockpiled supplies before the Volstead Act passed into law. It was rumoured that the Yale Club in New York City had stored enough liquor in its basement to keep it supplied for fourteen years. But for those without the resources and storage space, Prohibition was a serious imposition. Illegal drinking dens, known as speakeasies, and underground nightclubs mushroomed to meet public demand. Illicit alcohol, often brewed in unlikely and probably insanitary circumstances, was available to those prepared to pay for it, even if it did mean travelling to the shadier parts of town and muttering a password to some taciturn heavies on the door to gain entry. Huge profits could be made, and so organised crime took over the liquor business. There was also corruption: at Chumley's, one of the most popular New York speakeasies, the staff were told by the police that in the event of a raid they should usher their customers out of the exit leading on to Bedford Street, as the police would be coming in through the Pamela Court entrance. Purser Spedding of the *Aquitania*, a frequent visitor to New York in the 1920s, recalled the double standards that prevailed: 'I once attended a big police dinner in New York, and there were so many of the guests over the mark that the chairman of the feast called everybody to order, and said that if the drinking did not stop, he would send for a policeman. He had a whisky and soda in hand at the time, and this sally caused much merriment amongst the blue-uniformed guests.'[1]

American-owned liners were legally deemed to be American territory, so no alcohol could be served on them. However, it was unclear at first if the new regulations also applied to foreign-owned ships. Until informed that they were breaking the law, the many passenger liners docked alongside the wharfs of American harbours were unprecedentedly popular with thirsty visitors who, unable to get a drink on terra firma, were going aboard for dinner, dancing and cocktails, happy to pay for a night out.

Attempts to enforce Prohibition on other nations' vessels met firm resistance. Their protests were not solely on behalf of their passengers: on British ships there were the crew's rum rations to consider. The provision of a wine ration was not only legal but compulsory in some Latin ships. The Italian Lines pointed out indignantly that their officers, members of the crew and third-class passengers were entitled to a daily allocation of wine of not less than 12 per cent alcohol.

Eventually all foreign-owned ships agreed to keep their liquor stores sealed while in American waters, within three miles of the mainland; that boundary was later increased to twelve miles. Despite Prohibition, American cocktail barmen marked both occasions by creating, first, the Three Mile Limit (a heady mix of white rum, grenadine, cognac and lemon juice), succeeded by an even more potent brew, the Twelve Mile Limit (as before, but with a slug of rye whiskey). Thirsty passengers would make for the bars as soon as they departed New York, and a roaring trade commenced as soon as the ship reached international waters.

Paradoxically, in the same era as America was technically dry, cocktails boomed. Ocean liners prided themselves on the ingenuity and inventiveness of their bar staff; as British-born travel writer Basil Woon observed in 1926: 'The Atlantic has never been so wet as it has been since Prohibition started and Americans began travelling. What puzzles a ship's bartender is the baffling number of new cocktails the Americans have invented since 1919. Americans are no longer content to stay on a drinking diet of beer or whisky. They change their drink with every round.'[2]

The other major consequence of Prohibition for ocean-going ships was the inevitable temptation to smuggle illicit alcohol, for personal consumption or for sale. Between 1920 and 1933, Prohibition created an insatiable market for smuggled alcohol. Criminal gangs often approached the crews of ocean liners, offering them incentives to bring in large quantities of alcohol. A bottle of whisky in Britain cost about 12s 6d – the equivalent of 62 pence – and could be sold in New York for $5 – the equivalent of £1 – representing a decent profit. Demand was huge; young Americans carried hip flasks on nights out, containing any alcoholic tipple they could acquire. It was considered smart to drink alcohol, and even smarter to be able to acquire it.

The American gangs awaiting the arrival of the big liners might try to chisel down the price; seafarers might retaliate by watering the whisky. In addition, the Prohibition agents, who were employed to enforce the law, were often corrupt;

one of them was seen tottering back down the *Aquitania*'s gangway one evening with his bag so full of illicit booze he was attempting to bend the neck of a protruding bottle to conceal it. The captain and senior officers were expected to try to deter the practice by organising searches of the ship, but it is possible they occasionally turned a blind eye.

Violet Jessop had returned to work as a first-class stewardess with White Star Line in 1920, employed once again on the *Olympic*. Violet was indignant at the extraordinary lengths to which her passengers would go in order to smuggle drink ashore, especially as they often tried to involve her in their subterfuge:

> We were called upon as if it were our daily task, to help, advise, and often assist passengers to conceal their 'hooch' as we drew near to New York. It was all so fantastic. There were members of the Four Hundred, pillars of Wall Street, senators, lawyers, debutantes, card sharps, all with their minds on the same problem as we approached the shores of the United States. Under the circumstances, it was not surprising that often our men obliged, at a price, and even our women too. One ample-bosomed stewardess found she could carry off a quart of champagne in her 'balcony', and no customs official, however hard-boiled, had the nerve to tap the offending bottle with his little metal mallet used for such purposes, so she got away with it. Not quite so lucky was a small syndicate of stewards who ran quite a profitable business in liquor, until one day a coffin was needed in a hurry from the storeroom and their cache was discovered.[3]

Naturally, the absence of alcohol on American-owned ships presented no problem for teetotallers of any nationality, but it was an important factor in some passengers' choice of

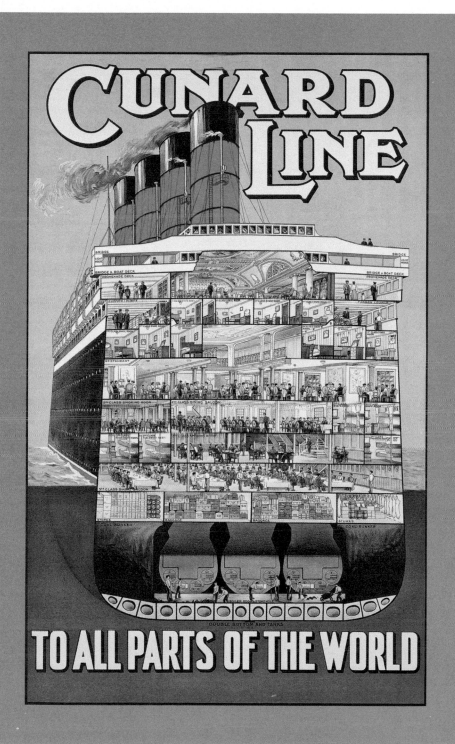

Cunard's 1914 poster showing a cross-section through the *Aquitania*.
The elite on the upper decks occupied luxurious staterooms. Second-class
cabins below were compact but comfortable. Down near the waterline,
third-class passengers slept in bunks and ate in functional dining rooms.

The Steerage by Alfred Stieglitz, 1907. Before the First World War, the poorest European migrants destined for America often travelled on ships notorious for their crowded and squalid lower decks.

Violet Jessop, 'The Unsinkable Stewardess', retrained as a Voluntary Aid Detachment nurse for the British Red Cross during the First World War. Surviving three maritime disasters, she worked at sea for more than four decades.

Marie Riffelmacher was fifteen years old in 1923, when she and two brothers emigrated from Hamburg to America, escaping the economic chaos of post-war Germany. The siblings' wages enabled other family members to join them in the New World.

Arthur Davis, co-architect of the London Ritz Hotel, designed the first-class lounge on the *Aquitania* in a similar style, recalling aristocratic country houses. The luxurious ship was dubbed the *Ritzonia*.

Refreshments were served on the open deck of the *Aquitania* when the weather allowed. Nervous passengers preferred not to dwell on the horrors of the *Titanic* disaster of 1912, but ample lifeboats were carried on the deck above.

Staunch Atlanticist Lady Astor sailed to New York on the *Olympic* in 1922. 'Gangplank Willies', reporters, photographers and newsreel cameramen surrounded her before she could disembark.

Hilda James, 'The English Comet', was an Olympic Silver Medal winner and sporting celebrity from Liverpool, who became a professional swimming coach on Cunard liners.

Swimming became popular in the 1920s and 30s due to an increased interest in health and exercise. Chic swimwear for fashionable women was now essential, and the first-class pool on the glamorous French ship, the *Normandie*, daringly allowed mixed bathing.

Formal evening dress was *de rigeur* for first-class transatlantic passengers, and well-heeled women travelled with different costumes for dinner and dancing every night, knowing they would be in the public eye.

Between the wars, organized deck games were a popular way of passing the voyage, and the prospect of winning prizes in front of an admiring crowd often brought out competitive instincts among female passengers.

Edith Sowerbutts, on the deck of the *Zeeland* in 1925, wearing the uniform she assembled for the new post of conductress.

Victoria Drummond MBE, Britain's first female seagoing marine engineer, pictured with anti-aircraft guns on *HMS Chrysanthemum* in March 1942.

Actress Tallulah Bankhead, and one of her many pet Pekinese, on her return voyage to New York in 1931 after eight years spent scandalising London society.

Josephine Baker from St Louis became famous in France for her extraordinary talent as a dancer and singer. Her dazzling stage and film performances captured the Zeitgeist of the interwar era.

Elsa Maxwell, self-taught party planner and impresario, followed the restless migratory habits of the international wealthy elite and established herself as their quintessential social 'fixer' on both sides of the Atlantic.

ships. Those who liked to drink were inclined to choose foreign-owned vessels, substantially reducing profits for the American companies. The strictures against the sale of alcohol on shore led directly to the invention of the 'booze cruise', developed to cater specifically for the determined party-goer. A brief maritime jaunt for a few days along the coastline of the North American continent to Havana or the West Indies in a foreign-owned ship provided an opportunity to party long and hard for the duration of the trip, in the company of like-minded revellers. Ostensibly this was for the purposes of tourism, but many pleasure-seeking passengers did not even disembark on reaching their supposed destination. What mattered was the opportunity for a luxurious blowout in a floating hotel with a hard liquor licence and obliging staff who would bring all manner of liquid refreshment to one's cabin, saloon table or recliner. Convivial cabin parties could be organised with the help of stewards, who were well-tipped to provide ample folding chairs and copious refreshments. The last night of a voyage before the ship re-entered American waters tended to generate a febrile atmosphere among the passengers:

> They were sailing towards the land of Prohibition, towards dry America, where no one knew how soon he would get a good drink again. So, everybody drank as much as he could and some of them more. Someone had made the statement, and it had become a belief, that all spiritous liquors would have to be poured into the ocean before the ship entered the harbour, so everybody tried his best to rob the tides of their precious booty … their weapons were the glasses in their hands and the empty bottles under the tables their trophies of victory.[4]

The enforcement of Prohibition in America also had unintended consequences for the transatlantic trade, providing an added incentive to travel – ostensibly to explore the world, but in fact to look for pleasure as well as business. This aspect was leaped upon by shipping companies, eager to fill their normally empty berths on the return trips to Europe, having landed the migrants who made up the bulk of their westward trade. With migrant numbers dropping due to stricter American immigration quotas, it made sense for shipping companies to upgrade their most economical accommodation, renaming it 'tourist class'. Much of the advertising and marketing of this era now focused on the affordability and sense of adventure involved in travelling the Atlantic. As both labour and fuel were then relatively cheap, steamships (which relied heavily on both) were comparatively inexpensive. Both the pound and the dollar were very strong against the post-war European currencies, so that a few dollars would buy excellent accommodation and wonderful food in France. The new tourists, American travellers on a budget, were increasingly attracted to the prospect of exploring mainland Europe. The Hemingways were able to live lavishly in France for $5 a day. In 1922 they dined at the best hotel in Kehl in Germany for 120 marks, the equivalent of 15 cents.

Long-distance sea travel had never been more affordable. The writer Alec Waugh, elder brother of the novelist Evelyn Waugh, undertook his first round-the-world trip in 1926. He claimed that it was more cost-efficient for him to travel by ocean liner than to rent an apartment in London. Thanks to his portable typewriter, he could work as a writer as well at sea as on land, selling travel features as he went. It was also possible to travel for years by writing about it. Writers carried their offices with them, and could work in isolation in their cabin, or in matey conviviality in a saloon. As a result, shipboard life started to feature in the literature of the era. P.G. Wodehouse's comic novel *The Girl on the Boat* perfectly captured the Zeitgeist, while Anita Loos's *Gentlemen Prefer Blondes* is largely set on board a transatlantic liner, and helped to establish ocean crossings in her readers' minds as the perfect environment for smart, modern young women on the make.

America and Hollywood beckoned to many British performers, entertainers and writers, keen to try their luck on the other side of the world and establish a career in the movies. Third-class cabin or tourist-class tickets were affordable for those living on frugal budgets, and it was a gamble that attracted many ambitious, creative women hoping to 'make it big' in the United States.

Elinor Glyn was a British-born novelist, and the younger sister of Lucy Duff-Gordon, who had survived the sinking of the *Titanic*. Elinor wrote romantic fiction to help support the finances of her husband, the spendthrift barrister Clayton. Elinor's novels sold reasonably well, but it was her notorious book *Three Weeks* that caused a sensation on publication in 1907. The heroine was an exotic Balkan queen, wealthy and worldly, who successfully seduced a much younger British aristocrat. Although it was a work of fiction, the book was rumoured to depict Elinor's affair with Lord Alistair Innes-Ker, younger brother of the Duke of Roxburghe, who was sixteen years her junior. In one scene in the book, the heroine lounges seductively on a tiger-skin rug, and this baroque detail piqued the interest of lusty widower George Curzon, who sent Elinor the skin of a tiger he had shot while Viceroy of India. They embarked on an affair, which lasted eight years and was the talk of London. When Elinor's husband obliged her by dying in 1915, she nurtured hopes that Lord Curzon might marry her once she was out of mourning. They seemed as close as ever, and the following year he asked her to supervise the refurbishment of an Elizabethan house he had acquired – Montacute, in Somerset. However, while she was working on the interior of the historic house alone in December 1916, she accidentally picked up a six-day-old copy of *The Times*. In the Announcements column, she read of the engagement of Lord Curzon to society beauty Grace Duggan. Elinor left Montacute, never to return, and condemned her former lover as 'so faithless, and so vile'.

Having acquired a public reputation as a passionate and promiscuous woman, in defiance of the social norms of the era, she bitterly resented the humiliation now inflicted on her by Curzon. A doggerel poem of the time ran:

> Would you like to sin
> With Elinor Glyn
> On a tiger skin?
> Or would you prefer
> To err with her
> On some other fur?

Elinor decided to escape hateful London society at the earliest opportunity. She was now fifty-two, recently widowed, and the object of public notoriety and scorn. However, her writing had earned her a decent living: *Three Weeks* had sold 2 million copies and had been translated into a number of languages. Still smarting from the end of the affair with Curzon, and confined to Britain by wartime travel restrictions, she nevertheless continued to support herself writing magazine articles and fiction, building up a considerable portfolio.

But America beckoned, and in 1919 she signed a contract

with Hearst's International Magazine Company to write stories and features for magazines such as *Cosmopolitan*. In 1920 she was offered a lucrative contract as a screenwriter for Famous Players-Lasky (which later became Paramount) and relocated to Hollywood. She became one of the best-known and most successful female screenwriters of the 1920s, and also produced and directed a number of movies. She was instrumental in establishing the careers of Rudolph Valentino and Gloria Swanson. Elinor Glyn's 1927 novel *It*, identifying the mysterious allure of sex appeal, was made into a successful film, and Clara Bow, the star, was known as the 'It' girl.

Elinor Glyn became a central figure in the English clique in Hollywood – a close friend of Charlie Chaplin, and of William Randolph Hearst and his mistress Marion Davies. She was influential as a film industry maven, advising Hollywood's professionals – both the on-screen 'talent' and the producers and directors who made the movies – on etiquette and deportment. She decided to return to England in 1929, aged sixty-five, to write fiction and her autobiography. The decade she had spent in America had successfully transformed her life, providing her with the chance to escape the censure and ridicule of one society, and to reinvent herself in Hollywood.

Diana Cooper was another British woman who hoped to make her fortune in the United States. In 1923, at the age of thirty, she went to act on the New York stage, for the very simple reason that she and her husband needed the money. As Lady Diana Manners, the daughter of the Duke of Rutland, she had been expected to marry an aristocrat. Instead she had fallen in love with Alfred Duff Cooper, an ambitious young chap from the Foreign Office, who had served honourably in the Great War and had been awarded the DSO. Their combined income on marriage in 1919 was £1,100 a year. They were hardly poor, but they needed substantial funds if he was to satisfy his ultimate ambition, which was to give up his civil service post in order to go into politics. In 1922 Duff Cooper devised the 'plan', which was for the two of them to go to America, where Diana, already an established performer, might make a fortune as an actress. It was a gamble, but if it was successful, he would seek a seat as a Member of Parliament.

Diana was a famous society beauty, and had appeared in two silent films. She attracted the attention of Max Reinhardt, an Austrian theatre producer, who needed actresses for a new version of *The Miracle*, which he proposed to put on in New York and then tour round the States. The play is a morality tale about a convent of nuns living in a medieval abbey that houses a life-size statue of the Virgin and Child, and the figure is believed to have miraculous powers. The starring roles are those of a young nun yearning for her freedom, and the 'living statue' herself, a physically demanding part as the actress had to stand immobile for nearly an hour, holding a heavy wooden baby, before apparently coming to life. Diana Cooper mostly played the statue, but would occasionally switch to the role of the nun.

Duff Cooper accompanied Diana to New York on the *Aquitania*, and he recorded: 'I enjoyed that journey as I have enjoyed all subsequent crossings of the Atlantic. We knew nobody on board, but we were sufficient to ourselves and for both of us the journey was a novelty.'[5] After six days he sailed home, for the first of their many long separations. He spent Christmas 1923 in the south of France with his mother, while Diana threw herself into rehearsals.

The production opened in New York in January 1924 and

was a spectacular success. Due to the increase in their income, the 'plan' was coming together. Duff Cooper made six transatlantic trips in ten months, before a general election was unexpectedly called in October 1924 and he was selected as the candidate for Oldham, near Manchester. He telegraphed Diana in New York urging her to sail the day after next on the *Homeric*. She was back within a week, and the couple successfully campaigned together. Duff Cooper was elected as an MP, Diana swiftly returned to the States, and the pair continued to commute across the Atlantic.

The Miracle was a hit for two winter seasons in New York, then it toured the States, with Diana still in the dual lead roles. This was followed by tours through central Europe, England and Scotland. The financial success of the play was vital to the couple's long-term hopes, but their only son, John Julius Norwich, also felt that the experience broadened his mother's horizons. For nearly seven years Diana lived in the world of Austrian-American-Jewish theatre, very different from British society, and she developed an affinity with that milieu. When the Second World War began, she sent her son to the States so that he might be safe in the event of a German invasion, and asked Dr Rudolf Kommer, Max Reinhardt's right-hand man, to act as the little boy's guardian.

Though emotionally devoted to his wife, Duff Cooper was notorious for his infidelities. John Julius Norwich reflected that, while the lengthy separations could have threatened their marriage, in fact they were both able to pursue their own interests for many months of each year. Diana had a fulfilling acting career, and her income enabled her husband to become a very distinguished politician. He was appointed British Ambassador in Paris immediately after the Second World War, a prestigious diplomatic role that they both greatly enjoyed.

Such Very Important Passengers were deliberately courted by competing shipping lines, in order to add *cachet* to their passenger lists. The glamorous Lord and Lady Mountbatten first sailed the Atlantic together on the *Majestic* in 1922 just after their marriage, and were allocated a sitting room, bedroom, wardroom, dressing room and bathroom, for the price of a single cabin. Their presence always generated excellent publicity for the White Star Line. The British public, who had avidly followed their on-off courtship and subsequent wedding, saw them as – in 'Dickie' Mountbatten's words – 'semi-royal'. It was only four years since the Armistice, and their nuptials were viewed favourably as a royal wedding that did not involve a German spouse, unlike the pattern before the war. The same public mood prevailed a year later when the second in line to the throne, Dickie's cousin Prince 'Bertie', married a small and pretty Scottish aristocrat called Elizabeth Bowes-Lyon, to general satisfaction.

During the voyage, Lady Edwina suffered from seasickness, and barely left their suite, but late at night Dickie would stroll up to the bridge for what he described as 'a companionable yarn' with the officers on watch. He took a great interest in the operation and navigation of the ship as he was a fully trained Royal Navy officer. When the newlywed Mountbattens arrived in New York, they were besieged by reporters keen to interview them, and 'snappers' wanting their photos for the papers. 'Simply ripping to be here,' offered Edwina gamely, in the manner of an Enid Blyton heroine. Next day they gave a press conference, then went to the cinema with Douglas Fairbanks and Mary Pickford. They dined with President Harding at the White House and attended the World Series at Madison Square Garden, where they shook hands with the baseball star Babe Ruth. After their east coast sojourn, a private railroad car took them to Hollywood, where they'd been lent the Fairbanks' house, Pickfair. Both were passionate movie fans, and they met Charlie Chaplin, who made a short film for them as a wedding present, casting the couple in a hold-up minidrama.

Leisure and pleasure aside, there were a number of pioneering women passengers for whom crossing the Atlantic in the 1920s and 1930s was an absolute necessity for their internationalist aims and professional activities. The formidable American-born Lady Nancy Astor, who had met her husband Waldorf on a transatlantic ship, had been elected the first woman Member of Parliament to take up a seat in 1919. Her parliamentary career as MP for a constituency in Plymouth, and her parallel roles as the mother of five children and a leading society hostess kept her busy in London and at the family home, Cliveden, in the immediate aftermath of the Great War. However, when she was invited to attend the 1922 Women's Pan-American Conference in Baltimore, Nancy took a seven-week leave of absence and set sail for New York with her husband on the *Olympic* on 13 April. Lady Astor did not go as an official representative of the British government, but she did go with the blessing of her political party, with a personal aim of promoting Anglo-American relations, to bring closer understanding between her adopted country and the nation where she was born.

While on board, Nancy occupied her spare time by exercising on the rowing machines in the ship's gym, and running circuits of the deck – unorthodox behaviour for a forty-three-year-old female politician in the 1920s. One of the New York newspapers had sent a female reporter on the transatlantic voyage to interview Nancy, and to transmit her articles about Nancy by radio while en route, a daring technical innovation. By the time they disembarked in New York on 18 April, Nancy Astor was a well-known figure in the city. The Astors were besieged by a vast number of journalists, and Nancy was inundated with requests for print and radio interviews. A celebrity on both sides of the Atlantic, there was American pride that a determined Virginian had blazed a trail into the heart of the mother of parliaments.

A staunch Atlanticist, Nancy spoke frequently and passionately on the need for future co-operation between Britain and America, but she kept away from the politically sensitive issue of Prohibition, though her pro-Temperance views were well known. The Astors met President Harding and were welcomed on to the floor of the Senate. They travelled by train to Nancy's birthplace, Danville, then headed north to Chicago and on to Ottawa, where Nancy addressed the Canadian House of Commons. There followed a busy schedule round the States, with many speaking engagements and a trip to Nancy's family in Virginia. Nancy's many frank statements had antagonised the Hearst press, particularly her trenchant views on the treatment of American forces' war veterans, and she was criticised in the many papers from Maine to California belonging to William Randolph Hearst, the millionaire newspaper proprietor.. Nevertheless, the tour was a triumph.

The Astors set sail on their return journey to Britain on the *Aquitania* on 23 May 1922. Waldorf's diary contains this brief last entry, and his relief is palpable: 'Final visits of friends. Final words with reporters. Final press photos. Ship sailed at 12 noon. After lunch I slept till dinner. After dinner I slept till breakfast the next day.'[6]

But Waldorf's hopes of a relaxing trip home were scuppered

by his vigorous wife. By chance, her formidable recent critic William Randolph Hearst was travelling on the same ship. Hearst was not keen on the Anglo-American 'special relationship', and was critical of many aspects of British life, though he preferred to travel on the *Aquitania* rather than any American-owned vessel. It was inevitable that when Hearst and Lady Astor met on board sparks flew. Purser Spedding tactfully recalled:

> Lord and Lady Astor are great favourites with the whole ship's company on board the *Aquitania*, and when they are travelling the ship is theirs ... to the great delight of both English and American passengers, also the stewards, who repeated to me as many of Lady Astor's remarks as they could remember. Nancy gave William Randolph the rounds of the kitchen, telling him exactly what she thought of him; the meeting occurred on deck one bright sunny morning in the presence of Mrs Hearst and her two sons.[7]

Throughout her long public career Lady Astor recognised the obligation to dress smartly and appropriately on all occasions, in order to be taken seriously in what was largely a man's world of politics. She had little intrinsic interest in clothes, but relied on the skills of her long-suffering maid Rose Harrison, who accompanied her on ocean voyages, as being on board ship was no excuse for lower sartorial standards. For women who were fashionable society figures, such as Lady Edwina Mountbatten and Lady Diana Cooper – both noted beauties – clothing and appearance were essential aspects of their public personae. For all wealthy women travelling in first class, any sea voyage required a large amount of clothing, as stylish passengers frequently changed from one outfit into another several times a day. Indeed, one appealing aspect of a transatlantic voyage for the modish was the prospect of showing off one's wardrobe and accessories over the course of six or seven days afloat, in evening outfits for dinner and dancing, and daytime clothes for all eventualities – breakfast in bed, lunchtime parties, sports attire for the gym, tennis court or pool, or stout tweeds for a walk on deck in autumn murk. In addition, passengers on the Atlantic run were advised to bring heavy outerwear such as coats for wearing on deck.

Despite the much trumpeted health-giving properties of exercise and bracing sea breezes, canny women travellers were wary about exposing their garments to ocean weather. Black wool could develop a green tinge because of the salt in the atmosphere, and the texture of serge could pucker if damp. Fur becomes heavy when wet, and can give off a distinctly animal odour. Feathers in hats and bandeaux would lose their *joie de vivre*, while metallic embroidery and sequins tarnished in salty spray. Notwithstanding modern depictions of inter-war heroines in evening wear glittering romantically at the ship's rail, the truth was more prosaic. Women who cared about their wardrobes stayed firmly indoors while sailing across the North Atlantic, especially when wearing evening dress.

For those travelling in first class, smart luggage was essential. In 1920 the London firm of Waring and Gillow ran advertisements to promote their brown cowhide suitcases, using the clumsy but unsettling slogan 'By their luggage you shall know whether they be well-born folk or not'. This phrase was guaranteed to play on passengers' insecurities, whatever their background.

It was important to look the part while travelling, and

vast quantities of luggage, from portmanteaux to hatboxes, dressing cases to hold-alls, added to the status of the traveller. Baggage contained clothes and matching accessories to wear on the voyage, as well as jewels, make-up and toiletries. Travel writers recommended Americans take one wardrobe trunk for a man and two for a woman; the second should be empty so that madame could fill it with clothes bought in Paris, or monsieur with Bond Street suits, spats and hats. Where possible, luggage should be stowed under the bunks or even in an adjacent cabin, to prevent these heavy objects being overturned and causing injuries during stormy weather. Any items 'not wanted on voyage' were stowed in the hold.

There was an impressive system for identifying and loading luggage of all kinds during the short turn-around time in port: every shipping line required each piece of luggage to have two labels attached, one tied on, the other stuck, detailing the passenger's name, address, ship and date of sailing. Cabin baggage would also be labelled with the number of the cabin and a large initial denoting the passenger's first name, so that it could be identified by the stewards. Hold luggage would be picked up by the company and delivered to the pier the day before sailing. The baggage master was in charge of sorting out the allocation of mounds of labelled luggage, despatching it to individual cabins. With porters and stewards to manage the 'heavy lifting', the well-heeled traveller would not see their luggage again until they arrived in their cabin.

Manoeuvring large cabin trunks aboard on embarkation was a fraught business; Violet Jessop described watching a hard-pressed steward grappling with one huge piece of luggage, mounted on a hand-barrow:

Down the long alleyway a huge wardrobe trunk on its truck moved waveringly forward. It was the kind of wardrobe trunk without which, at that period, no American woman would think of travelling. It hesitated, as passengers and friends crossed the path of its uncertain progress: now and then, around one side would peer a face – perspiring and apparently sorely tried – but equally determined to carry on.[8]

A number of specialist firms provided indispensable trunks to suit the need of every elegant traveller. The Parisian company of Louis Vuitton was renowned for its ingeniously designed, practical and chic luggage. Reassuringly expensive, every piece was covered in the company's distinctive 'monogram' patterned canvas, featuring the founder's initials. One popular design for sea passengers was a large desk trunk, which resembled a tall, compact chest of drawers. It took up comparatively little space in cabins, but its many compartments neatly held accessories and small items of linen.

Packing for a sea voyage was an intricate business. Of course, the wealthy passenger would not pack their own luggage before departure. That was still a specialist job for a servant, though a steward or stewardess might be pressed into service to unpack when the luggage was deposited in the cabin. 'With expedition and with a sure hand, because her task is accomplished without anxiety, the maid can pack the trunks and bags that will be shipped across the Atlantic, for fashion is now extremely kind to the traveler.'[9]

Many of the better-off first-class travellers brought their own maids or valets along on the voyage specifically to care for their wardrobes and general appearance. Body servants, as they were called, were usually accommodated in special interior cabins linked to the individual passenger's stateroom or cabin, so as to be constantly available for their employers. On the luxurious *Olympic*, the servants shared a special dining room for maids and valets, and exchanged gossip about their employers: 'Mary Pickford's maid tells Rudoph Valentino's valet how many pairs of silk stockings Mary has, and Rudolf's valet tells Mary's maid about Rudolf's favourite purple satin pyjamas.'[10]

The printed passenger list for a typical trip on *Aquitania*, leaving Southampton for New York on 25 June 1921, lists alphabetically the names of 464 people travelling in saloon class, as first class was known on this ship. Nineteen of them had brought their maids, eight had their valets, two had both a maid and a manservant in attendance, and one person was accompanied by a nurse. Second class listed 370 passengers' names, but no servants. Those travelling without servants inevitably relied on the assistance of their cabin steward or stewardess, and tipped accordingly. The most fashionable ships catered for their passengers' concern with their appearance by providing fine laundering aboard, and an ever increasing workforce of dressmakers, masseuses and beauty therapists were employed to improve their female passengers' sense of wellbeing and bonhomie. Many of the women engaged by the shipping firms to pamper their female passengers originally came from quite ordinary or humble working-class backgrounds, and often grew up in port cities such as Liverpool or Southampton. The positions newly available on board the great liners ploughing across the Atlantic provided excellent opportunities for them to earn a good living, to travel the world, and to acquire a level of sophistication and worldly knowledge that would not have been available to them had they stayed on dry land.

Every liner with any pretensions to passenger comfort carried a number of female hairdressers to meet the constant demand. Between the wars, maintaining an elegant and *soignée* appearance while afloat was considered essential for any woman with a sense of style, but even the most chic coiffure could be turned into a frizzy mess, thanks to the damp, wind and salt spray. While the wealthiest women travelled with their own maids, who could dress the mistress's hair every day, most would rely on the dexterity and skilled expertise of the ship's hairdressers.

The mother of a future Archbishop of Canterbury was known as 'Madame Edna', a professional hairdresser who made home visits around Liverpool. Ann Benson had married a Scottish-born engineer, and they lived in Crosby with their four children. The three eldest were close in age, but the youngest, Robert, was a much later addition to the family; he described himself as an 'autumn leaf'. Ann was always keen to travel, and Robert recalled a family friend, a Cunard steward called Bill Barnard, who would bring them gifts of exotic American chocolates and cigarettes obtained on his voyages. Ann, who was rather a volatile beauty with large brown eyes and curly auburn hair, found home life both humdrum and stifling, and suffered badly from *ennui*. She went to work as a freelance hairdresser on a succession of Cunard passenger ships in the early 1920s, and chose the more fashionable name 'Nancy' as her on-board professional persona, 'Madame Edna' being perhaps a little too ponderous. Nancy Runcie and her friend Peggy Levy signed up for a world cruise with Cunard in 1923. Two-year-old Robert was handed over to his great-aunt, while the rest of the family relied on their easy-going and sociable father, the capable eldest sister, Kathleen, who was fourteen, and the help of a servant. Ann's voyage took her to San Francisco, Cairo and Yokohama. Robert's older brother, Kenneth, who was ten at the time, remembered feeling forsaken because their mother was away for nearly five months, but the other siblings accepted that their father stayed close to home while their mother travelled overseas. Although unusual by the general standards of the 1920s, for the families of seafarers of both sexes parental absences of varying lengths were a matter of routine, even necessity. Seafaring mothers often relied on their extended families – or even neighbours – to look after their children on an informal basis while they were away at sea, sometimes for months at a time, and small but useful sums would change hands to recompense the children's adoptive 'aunties'.

Ann Runcie was employed on sea voyages where demand was high for lady hairdressers (as opposed to ladies' hairdressers, who were men). She usually worked freelance at home, so could set her own rate for different services. However, if they were engaged by the shipping company, lady hairdressers were paid at the same rate as the ship's assistant barber. This was a great improvement on the wages usually available to them ashore, and there were also tips from grateful passengers. While Ann's earnings usefully augmented the family's finances, it seems her contact with affluent passengers on board ship also made her extremely ambitious for her offspring. Ann had had a lacklustre education, but was a keen reader. She encouraged young Robert to study the classics at school, an unorthodox decision in Liverpool in the 1930s – because of the Depression, most families wanted their children settled in a steady job, or an apprenticeship, as soon as they could leave school. However, Ann was thrilled when Robert gained a scholarship to Christ

Church, Oxford. He went on to serve in the Scots Guards, renowned as a 'smart' regiment; she would pump him for names of his fellow officers, then go to the public library to look them up in *Debrett's*. Tellingly, Robert Runcie later described himself as a 'chameleon', with an ability to adopt the accents of his social group wherever he found himself.[11] He lost his Merseyside accent and took on the voices of his public-school friends, first at Oxford and, later, in the army, a social skill that he may have inherited from his observant, aspirational and hard-working mother.

By the mid-1920s the range of roles available to women seafarers on the ocean liners was expanding, and there were now unparalleled career opportunities for many women from less privileged backgrounds. The long-standing assumption that working women would be tolerated by their male colleagues only in 'nurturing' capacities, as stewardesses or nurses, was being eroded. Any shipping line enterprising enough to provide novel and amusing daytime diversions for its wealthy but occasionally bored lady passengers would be likely to improve its reputation and ensure their repeat custom.

One new role for women seafarers was that of swimming instructress. The post-war obsession with health, physical fitness and the enjoyment of sport, known as the Cult of the Body, had made it fashionable for more liberated women to swim for pleasure as well as for exercise, and the transatlantic ships had impressive heated seawater indoor pools for passengers' use. Women and children swam at separate times from the men, so female pool supervisors and professional swimming instructresses were required to ensure their safety, and to provide lessons and demonstrations of diving and swimming techniques on request. Demand was high: Mrs Nan

Palmer spent six years as swimming instructress on the *Majestic*, employed by White Star. She claimed that on one particularly hectic voyage she spent two whole days in the pool, except for mealtimes.[12]

But it was Hilda James who was the trailblazer in this field. Born in 1904, Hilda was a window cleaner's daughter from Liverpool, and a remarkably talented swimmer. She won a silver medal for Britain at the Antwerp Olympics in 1920, aged only sixteen. There she met the American Olympic team and became friends with their manager, Charlotte Epstein, captain of the Women's Swimming Association, based in New York. They taught her the new swimming stroke known as the American crawl, and she was subsequently unbeatable, setting five new English records and two world records in just three months in 1920. Throughout 1921 swimming records continued to fall to Hilda, who was known as the English Comet. She became a popular sporting celebrity, and her triumphs were much featured in the press, especially in Liverpool.

In 1922 the Cunard company offered to provide free passage for Hilda to travel to the United States to take part in invitation swimming races, galas and exhibitions. Hilda's mother, Gertie James, grudgingly agreed to accompany her as chaperone. In addition, the company gave Hilda a complimentary life membership of the Cunard Club at the Adelphi Hotel in Liverpool, so she could practise swimming in the pool. The company was sponsoring one of the country's top sporting personalities, someone who they regarded as a public relations asset, and a potential future employee.

Hilda, her mother, her coach Mr Howcroft and his wife Agnes, travelled by train for Southampton on 21 July 1922. They were provided with second-class cabins on the *Aquitania*, but were entitled to dine in the first-class Louis XVI dining room. They were invited on to the bridge, and given an escorted tour of the ship. Hilda proved to be very popular with the passengers, the captain and the crew. She gave a number of swimming demonstrations in the indoor pool, and she coached swimmers, which she enjoyed.

On arrival in New York on 28 July 1922, Hilda was amazed by the vibrancy of the city. With her American friend Charlotte (known as Eppie), she enjoyed a formal dinner at the Waldorf Astoria, went sightseeing in New York, took a trip up the fifty-seven floors of the Woolworth Building, marvelled at the Statue of Liberty, rode on the Staten Island Ferry, and visited Central Park.

Touring the north-eastern States was an education for Hilda, who was still only eighteen. The tour was intended to raise the public's awareness of swimming as a suitable sport for women. Various clubs held galas and invitation events, and the local competition against Hilda was fierce. She broke a number of British and American records, and at Indianapolis she watched a young champion swimmer competing. A tentative friendship developed between them; his name was Johnny Weissmuller.

After returning to New York, Hilda and Agnes Howcroft were escorted by Eppie to Bloomingdale's, the smart department store, to try on frocks, as there was to be a formal dinner-dance at Madison Square Garden to mark the end of Hilda's tour. Hilda and Agnes enjoyed the fantasy of dressing up in evening gowns, but knew they could not afford to buy anything to wear. To their amazement Eppie presented them with dress bags containing the lovely outfits they had just tried on, and with matching shoes and handbags. The clothes had been paid for by various American swimming organisations and benefactors. Due to Hilda's amateur status, she could accept 'gifts in kind', though not cash. Still reeling from their benefactors' generosity, Hilda and Agnes were given a professional makeover. Hilda's humble background and teenage years spent competing in swimming tournaments had not accustomed her to beauty treatments, but the results were impressive. Hilda and Johnny Weissmuller were the joint guests of honour at the dinner; Johnny danced with Hilda and gave her the medal he had won in Indianapolis the first night they had met.

Hilda escaped her chaperones to spend a romantic day at Coney Island with Johnny, and they returned to Manhattan without Hilda's truancy being discovered. Johnny was still painfully shy, but they exchanged a kiss. Weissmuller's swimming career was just taking off: he went on to win five Olympic gold medals, and broke nearly seventy world records. Handsome and with an impressive physique, he became a model, then an international film star, playing Tarzan, King of the Jungle, in twelve movies.

Hilda, her mother and the Howcrofts returned on the *Mauretania* on 5 September – they had been upgraded to first class. As before, Hilda gave swimming shows and coached her fellow passengers. At the captain's table one evening, she was told that Cunard was planning a new type of ship for ocean cruising, that it would want swimming instructresses, but also intelligent, sociable 'people people'; had she thought of turning professional? Hilda was tempted, but in order to compete in the 1924 Paris Olympics she needed to retain her amateur status, so she declined.

A rapturous reception back in Liverpool was hosted by the city's Lord Mayor. By crossing the Atlantic, Hilda's life had been transformed. She had adapted to the social mores of life afloat, and impressed her fellow passengers and the officers with her affability. Her experiences in America had shown her that a more glamorous life was possible, and she had made genuine friendships. She planned to move to New York when she was twenty-one; however, she was still three years away from living independently, and her controlling parents increasingly viewed her achievements with jaundiced eyes.

In September 1922 Hilda set another world record, for the 150 yards freestyle, and began training in earnest for the British Olympic team. She was a strong contender, and was predicted to secure three gold medals, despite her youth. But in November 1923 a huge family row erupted in the James household. Hilda's mother Gertie, always prone to furious and irrational rages, refused to let her daughter join the British team abroad unless she could go too. The Paris Olympics were scheduled from 4 May until 27 July 1924, and the Olympic Women's team was provided with official chaperones. Mrs James could travel with them, but she would have to do so at her own expense. Gertie felt entitled to benefit from her daughter's sporting prowess. If she wasn't offered a complimentary trip to Paris, she would not permit Hilda to participate.

What had started as a heated family argument escalated into an ugly and horrifying assault. Hilda's father John had a nasty temper; goaded by his histrionic wife and stung by his daughter's defiance, he gave Hilda a savage beating with his leather belt, leaving her unconscious. The following morning Hilda could barely stand; she dressed with great difficulty, packed a small case, and escaped to her Aunt Marjorie and Uncle Jim's house nearby. Her relatives were appalled, and called a doctor who treated her injuries. It was three weeks before her wounds began to heal. Aunt Marjorie berated her brother, and warned him that they had informed the police of his assault, to deter any future violence.

Hilda, badly bruised and traumatised, was denied the chance of participating in the 1924 Olympics in Paris by her parents. Having produced a daughter with astonishing talent, dedication and great willpower, they had deliberately sabotaged her chances, just as she was on the brink of international success. Hilda was now counting the days till her twenty-first birthday, 27 April 1925, when she would be free to live independently. Having missed selection for the Olympics, it was no longer necessary to maintain her precious amateur status. She could turn professional, carry on swimming and make a living from her talent.

Cunard offered Hilda a job as the 'resident professional' at their swimming club at the Adelphi Hotel. Yet again her parents were implacably opposed to any professional career for her and refused to give their permission. Hilda didn't risk another confrontation and possible punishment beating. This time she would use subterfuge and rely on the support of friends. She replied to Cunard, asking if the company might be prepared to employ her once she was twenty-one. Sir Percy Bates, deputy chairman, wrote back, assuring her the company would wait, and invited her to tea with him and his wife at his home, Anderton Hall.

Sir Percy explained to Hilda his plans to develop a new series of liners for leisure cruising. He wanted Hilda to travel on board as an employee and give swimming displays and lessons to the passengers. She was a celebrity, and she had proved herself to be entertaining company, sociable and resourceful. He also gave her his telephone number, 'in case there was any more family trouble at home'. Hilda realised he probably knew about her beating, as he was a local Justice of the Peace and had good contacts with the police. Percy wanted to offer his support in a practical way, with a job on Cunard's new cruise ship, the RMS *Carinthia*, which was to be operational by the summer of 1925. For Hilda this was the opportunity that changed her life.

The *Carinthia* was designed as Cunard's cruise liner, running between Britain and New York, with world tours every winter, and shorter summer cruises. Percy wanted Hilda as part of his hand-picked team of senior officers. Officially she would be the swimming instructress, with regular pool and swimming duties, but also act as an entertainment hostess to the guests as required. This was a new role – Cunard needed someone to charm the passengers, helping to keep them amused. Hilda accepted, but on condition that her new job could be kept secret from her family till the very last minute in order to avoid another family row.

Hilda was discreetly put on the Cunard payroll on 1 July 1925. She was given a secret tour of the *Carinthia*, during which she changed into her swimming costume and posed for publicity photos on the edge of the pool. She gradually smuggled her few belongings out of the family home and stored them in the Adelphi Hotel. On the morning of the *Carinthia*'s maiden voyage she planned to set out for central Liverpool as though she was going to the Adelphi for a swim, leaving a letter for her father to find after the ship had sailed.

But the day before Hilda's planned escape, there was another volcanic family row, and Hilda finally snapped. She told her mother that as she was now twenty-one years old, she was independent and that the next night she would be leaving Liverpool on the new Cunard ship the *Carinthia*. Gertie was inarticulate with fury, and, screaming uncontrollably, she flung a metal pan at Hilda, who dodged it. John heard the commotion and came running into the kitchen. Hilda slipped out of the door before he could stop her, and escaped to her aunt and uncle's house once more.

The following day, Saturday, 22 August 1925, there was a traditional maiden voyage send-off from the Pier Head at Liverpool. Brass bands oompahed as passengers embarked on the *Carinthia*, the most modern and luxurious of Cunard's ships. Luggage was loaded, photographers and reporters recorded the occasion, and there was a general air of anticipation and celebration.

On the packed quayside, her sister Elsie appeared, carrying the modest suitcase Hilda had left behind in her flight the night before. The two sisters hurriedly exchanged news. After Hilda's escape, Gertie screamed so much that she made herself sick, and then sat in a chair seething for the rest of the evening. John James had hurtled out of the house in pursuit of his errant daughter. He returned eventually in a towering rage, prompting Gertie to start ranting about hell; the resulting row between the parents continued unabated throughout the night. Elsie and their two brothers had hidden in fear. The following morning, John instructed Elsie to collect together all Hilda's remaining belongings and dispose of them. When Gertie started declaiming on her favourite theme of sin, John told her forcefully that he wished that she had left instead of Hilda, and thundered out again.

That day's copy of the *Liverpool Evening Post* prominently featured the publicity photos of Hilda poised on the edge of the *Carinthia*'s pool, and the announcement of her appointment. Elsie knew that when the paper was delivered to the James family home in the afternoon it would lead to further drama, so she had collected Hilda's remaining few possessions and brought them to the Pier Head, together with Hilda's small suitcase. In addition, she gave Hilda a thick envelope and told her not to open it till safely at sea. The sisters agreed to write to each other using a go-between; Elsie wished her luck and Hilda went aboard in tears.

Hilda's twin cabin was in the female-only crew quarters, near the medical room. She was to share with one of the hairdressers, and the luggage she had hidden at the Adelphi had already been delivered to the cabin. Their corridor was lined with twin rooms – the other female crew were cleaners, laundry staff, nurses, stewardesses and a physiotherapist, but Hilda was the first swimming instructress.

At 5.30 p.m. Captain Diggle gave the order and the *Carinthia* inched out of Liverpool on its maiden voyage. Onlookers and passengers cheered, brass bands played, hats were waved and an exuberant send-off was enjoyed by all. There was a salute from other craft on the Mersey, a chorus of whistles, hoots and toots, a venerable tradition for a first voyage.

Elsie returned doggedly to the festering family home, while Hilda hid in her cabin until winkled out by a new female colleague, and taken off to have supper in the crew's dining room. Later, Hilda had a crisis of confidence. She had left behind everything familiar, and severed her links with her parents. She had no idea what the new job entailed, or whether she would be able to do it. In the envelope Elsie had given her she found a letter: her sister promised to look after their younger brothers, and there was some money, which she had saved from her small teaching salary. Lonely and exhausted, Hilda cried herself to sleep.

The following day was better. The other women on board were thrilled to discover they had a celebrity, an Olympic medal-winner, as their new workmate. Hilda was given a programme including a timetable for coaching passengers in the luxurious pool, which was designed in Roman style with marble columns, seating, a sauna and well-appointed changing rooms. There were curved staircases down from the deck above, providing a *grande descente* for the sporty, a witty reference to the elegant and fashionable staircase above decks in the restaurant. Hilda had a small office too, next to the pool, which offered a welcome escape from her tiny cabin. Passengers flocked to the pool with requests for personal tuition.

She was invited to dine one evening with Captain Diggle in the Adams Room, in first class. Fortunately, she had the gala dress bought for her at Bloomingdale's in 1922, which she had smuggled out to the Adelphi prior to her escape. The seamstress insisted she must wear it, the laundrymaid pressed it for her, and the hairdresser tackled her hair. Hilda was also initiated into the mysteries of applying make-up, a novelty as she had spent most of her formative years plunging into or thrashing around in cold water. In addition, she had never tried an alcoholic drink before – like so much of her new life, her parents thought it 'sinful'. At the captain's table she was seated next to the ship's senior wireless officer, Hugh McAllister, whom she had first met on her trip to New York. They were delighted to see each other again; he had known for months that Cunard intended to employ Hilda on the *Carinthia*, and he had signed on in the hope of meeting her once more.

The ship arrived in New York on 31 August 1925, and Hilda had five days' shore leave. She met her old swimming pal Eppie who knew the true story behind the 'lost Olympics'. Eppie was convinced that if Hilda had competed in Paris the previous year, the American team would not have swept the board of medals as they had.

A newly confident Hilda wrote a proposed job description for herself, as requested, and sent it to Sir Percy. She wanted to be more involved with the passengers beyond the confines of the pool, by organising deck games and devising on-board competitions to entertain them. Sir Percy agreed, gave her a new title – Cruise Hostess – and raised her salary substantially. In future she was to travel in a second-class stateroom – a vastly superior billet – and she was also provided with a set of formal uniforms, to match the male officers' 'whites'.

For the next four years Hilda worked as a Cruise Hostess on board Cunard's new fleet of luxury liners, accompanying exclusive world cruises organised by American travel agents Raymond-Whitcomb. The itinerary varied, but the wealthy clientele were mostly Americans, industrialists and financiers, theatre stars and millionaires, and they would be travelling for months. Because of the strictures of Prohibition, the bars on board had to wait until the *Carinthia* left US territorial waters before they could open and there ensued an all-night party.

Hilda's prime role as hostess was to entertain the passengers, to get to know them and find ways to keep them amused, for which she had to be adaptable and resourceful. There were formal evening dances, with music provided by the band, but Hilda and the gym staff also devised daytime pursuits, deck games or board games. She learned card games and developed a number of casino tricks, such as impressive card shuffles. She had bought a book of crossword puzzles in New York and had copies printed up by the on-board newspaper office. She created treasure hunts, and staged crimes that had to be detected by the passengers. She also led the swimming activities, coaching individuals, and organising water polo or water volleyball. The pool became a centre of fun and jollity.

Hugh McAllister was very attentive, taking her round the wireless office and the bridge, and inviting her to the officers' mess, which was unusual for any female crew member. She hosted an out-of-hours pool party for officers, by way of return, and unsuccessfully attempted to teach Hugh to swim.

By signing on for all available work as a cruise hostess, Hilda managed to avoid seeing her parents for a year, making a good living crossing the Atlantic and touring the world, her salary augmented by generous tips from her passengers. Her loyal siblings Elsie, Jack and Walter managed to negotiate a partial reconciliation with her parents, so, weeks later, Hilda turned up at the family home dressed in a leather coat and astride a brand-new motorcycle. Her parents were appalled, but Hilda didn't care, and took her siblings out riding pillion. The motorbike brought her even more independence, as she could travel quickly around Liverpool whenever she was back in port. In 1926 Hilda received a 'very strongly worded letter' from Nancy Astor MP, who had been visiting the city and was walking along the promenade at Parkgate when Hilda thundered past. Lady Astor wrote to Hilda criticising such unladylike behaviour from a supposed role model for the younger generation. Hilda took no notice, and, tellingly, Nancy Astor – a speed enthusiast – acquired and learned to ride her own motorbike during the Second World War, when she was nearly sixty.

Hilda's life was transformed by her new status and her confidence. Whenever she performed on land she was always billed as 'the Cunard World Champion Swimmer Hilda James'. She completed a number of world cruises, survived hurricanes off the coast of Bermuda, and visited Hollywood, Hawaii, Cairo, Australia and the Baltic. Daringly, she would smoke the occasional cigarette, and have a beer or two while dealing the cards in the officers' mess. On board the orchestra played the latest dance tunes, and Hilda could Charleston.

By now Hilda and Hugh were definitely a couple. Hugh was no Johnny Weissmuller; he was a poor dancer, and he had no talent for swimming. But sitting together on a trip ashore, he held her hand for the first time after six years of friendship. In 1929 Hugh proposed marriage, a development that surprised none of their acquaintances, and a celebratory engagement party was held on board the *Carinthia* in New York. Sir Percy accepted Hilda's resignation when their ship returned to Liverpool. She was now to become a professional swimming instructress in Liverpool, and her sailing days were over, though Hugh continued going to sea as a wireless officer for Cunard. They married in September 1930, despite sullen resistance from Hilda's parents, and their son Donald was born the following May.

Hilda James travelled the world first as a passenger, invited to America to compete as a talented sportswoman, and later as a seafaring professional. Working in international travel in the 1920s allowed her to see how different life could be, and provided opportunities to transcend her modest beginnings. She was helped by the generosity and goodwill of her many friends, and gained the confidence that might have eluded her if she had stayed at home, cowed by her parents. Like many women from less well-off backgrounds, sailing the Atlantic as a career transformed her life.

For many of Hilda's female contemporaries, the great ocean liners offered career opportunities unimaginable before the Great War. Administrative roles, such as stenography, were first opened to women as a direct result of the wartime shortage of male labour; now it was possible to gain employment as a typist in the purser's office, or to travel as a private secretary to a wealthy international passenger. Hairdressers such as Ann Runcie could change their professional names to something more aspirational, leave their children to be looked after by family members and sail away for months at a time to satisfy their wanderlust, augment the family income and raise their social aspirations by filling well-paid positions on Cunard liners. The seasoned stewardess, the 'unsinkable' Violet Jessop, was recruited to take to the seas again, her hard-won experience deemed invaluable in meeting the increasing demand from female passengers.

On the North Atlantic run, their clients were predominantly the women of the upper decks, the privileged and fashionable, such as Lady Mountbatten; the creative, such as Elinor Glyn. There were celebrities and performers, such as Lady Diana Cooper, who flitted across the ocean balancing her career commitments with those of her family. There were international figures intent on improving Anglo-American relations, such as Lady Astor, who were travelling at their own agency, in pursuit of their own goals, even if they did occasionally have their respective husbands 'in tow'. After four long years of war, women of means were demonstrating an independence of action and movement, and as the golden age of transatlantic travel dawned, they required unprecedented levels of services from the on-board female workforce.

5

Edith and Her Contemporaries

Edith Sowerbutts surveyed her reflection critically in the full-length looking glass. She had planned to buy the correct uniform from the Nurses' Outfitting Shop in Victoria Street, London, but they did not yet stock what she required, as her job had only recently been created. Resourceful Edith had therefore invested in a lady's greatcoat, standard issue for Canadian-Pacific's stewardesses, and had found a suitably smart navy blue frock-coat with scarlet piping in an Oxford Street department store. The addition of some brass buttons had given her improvised uniform a more maritime look, an impression augmented by a neat navy blue hat. Her pin-on metal badge read 'Conductress' in gold letters, and she was pleased with her new persona.

It was the summer of 1925 and bespectacled, freckled Edith, aged twenty-nine, full of energy and possessed of a hearty appetite, had been recruited by Red Star Line to be a conductress on their ships travelling between Antwerp and Canada. Her role was to look after the welfare of unaccompanied women and children emigrating to Canada, especially those deemed vulnerable to possible exploitation.

While the United States restricted the influx of immigrants in the 1920s, Canada actively recruited Europeans, promising them ample employment opportunities. The country needed domestic servants, cooks, waitresses, teachers and nurses, and financial assistance was available for suitable applicants. To meet the demand, in 1924 Cunard put two of their passenger ships, the *Caronia* and *Carmania*, on the run from Liverpool to Canada, via Belfast, while Red Star Line ran regular ships from Antwerp to Halifax and New York, catering primarily for continental émigrés.

Single women contemplating travelling to Canada were reassured that, for the first time, there would be professional female chaperones on board. There had been considerable international concern about the 'white slave trade', the trafficking of women and children. On arrival in a vast city where they did not speak the language, solo female passengers could be easy prey for the unscrupulous, and tales abounded of undocumented and unaccompanied victims being lured or coerced into the sex industry. The Canadian government insisted that immigrants heading to their shores should be accompanied by professional, competent welfare officers, called 'conductresses'. Each passenger's biographical details and their emigration plans would be recorded while they were on board the ship, and after being handed over safely to the proper authorities on arrival in Nova Scotia their onward journey and eventual settlement would be monitored.

The creation of the new role of conductress provided British seafaring women with their first increase in status, to the rank of officers. Conductresses were competent, authoritative women, who commanded respect within the ship's company while chaperoning their charges. Unlike stewardesses, who provided practical care for the physical comfort of their allocated passengers while afloat, conductresses were primarily responsible for their passengers' moral welfare. Conductresses escorted, advised and protected the women and children on board, especially (though not exclusively) those in third class, who were crossing the globe, travelling to an unknown future in a country where they did not yet speak the language.

Edith Sowerbutts was one of a small but influential group of women who travelled the North Atlantic as conductresses. She started working for Red Star Line, which was owned by White Star, in 1925, and continued in that role for six years, when assisted immigration to Canada ceased and she was made redundant, although that was not the end of her maritime career. Edith was well-travelled, experienced and adventurous. In 1919 she and her friend Trix Bickerton, a former suffragette, had worked their three-month passage as stewardesses on the *Canberra*, a troopship returning exuberant demobbed Australian soldiers to Sydney. The experience was enlightening; on arrival Edith and Trix were described as 'A couple of bonzer Sheilas, but no bloody good to me as stewardesses' by the chief steward. Attracted to the vibrant life in Sydney, Edith stayed in Australia for several years. She had trained as a stenographer, and although she always disliked typing, it was a marketable skill that brought her plenty of work. She bought her first typewriter, a second-hand Underwood, for £20, and in her spare time she wrote articles for the *Sydney Morning Herald*. She recalled:

> I have often pondered the question: was the typewriter women's road to equality? I think maybe it was one of the first steps. Myself, I could see very little attractive in employment where one typed away from morn to eve, but it has been and can still be a means to an end. I escaped from the typing pool soon after war was declared in 1914. I found work with more scope, less typing, even less shorthand.[1]

Edith returned to Britain by sailing the Pacific and crossing Canada by train. She was employed by the Society for the Overseas Settlement of British Women (SOSBW) to promote overseas migration, and ran the organisation's stand at the 1924 British Empire Exhibition in Wembley, during one of the chilliest and wettest summers on record, a factor that may have helped recruitment. She had a great deal of practical experience of intercontinental travel, as well as fellow feeling for people who were prepared to travel to better their lives. Edith was also a natural champion of those who were discriminated against on the grounds of race, class or gender. Her innate sense of justice made her a formidable advocate on behalf of the passengers in her care, and she was delighted to be taken on as a conductress by Red Star Line in 1925.

The westbound route of Edith's first ship, the *Zeeland*, was from Antwerp, via Southampton to Halifax in Canada, and then to New York. The majority of people carried in third class were would-be emigrants from all over Europe, planning to settle in the New World. Edith received £12 a month; by comparison, an assistant purser on a small liner would get £15 a month, while a ship's doctor would have a basic salary of £30–£40 a month, and charge additional fees for any services. Male officers had an entertainment allowance on top of their salaries, so that they could 'treat' passengers to drinks at the bar, but this was not given to conductresses. Edith was often short of money as she was fond of the high life once ashore.

Conductresses were expected to 'head' a dining table every evening in the first-class dining room, acting as hostesses for any unaccompanied women. On her first voyage out on the *Zeeland* Edith hosted a table of ten elderly American ladies, who seemed to find her presence reassuring. Edith was told by the purser that she would be expected to change for dinner, 'as you would at home', which amused her, as her family was quite ordinary and did not change for their evening meal, a simple supper. She owned a neat little black dress for evening wear, and a couple of white *piqué* sleeveless tennis dresses for her off-duty hours on the sports deck in the summer months.

While Edith was available to unaccompanied women in all classes, her primary role was looking after the interests of those in third class, and processing their immigration applications. She would introduce herself to each one, explaining that she had a list of official questions, and record the answers by hand, then type up the details later. On each voyage Edith compiled detailed lists of all unaccompanied women immigrants across all three classes for the Canadian authorities. To extract this information from each woman was often a race against time, because third-class passengers were housed in the least stable section of the ship, and therefore prone to seasickness. If possible, she completed her interviews before they passed the west coast of Ireland, after which the open Atlantic was rougher. Edith relied on the services of interpreters, and was particularly fond of a remarkable character called Terps, an Orthodox Jew who spoke fourteen languages. He spent most of his free time in the kosher kitchen with his friend the chef. Edith used to join them there for fish and chips, as she thought it was the best food available on the ship.

Edith had a great deal of sympathy for her passengers, and believed that, whatever their previous experiences, they were heading for a better life in Canada or America. The unaccompanied women in third class were typically Poles, Ukrainians, Yugoslavs, Greeks, Italians, Romanians or

Germans. Many of the central Europeans were from poor rural backgrounds, and they had already made a mighty journey, usually by train, across Europe to get to Antwerp to join the ship: 'They wore long, voluminous skirts or dresses, grubby but oft-times hand-embroidered; short sheepskin jackets, head scarves and high boots. Their hand-worked blouses were made of a coarse fabric resembling calico. They wore their hair in plaits. I had never seen their like before … these people had known nothing but a very hard life – mud floors, no mod cons.'[2]

The voyages made by individual women to the New World often required great stamina and determination, and those making the journey were not always young and fit. In 1926 *White Star Magazine* carried a brief article about a nonagenarian who had made an epic journey:

> A remarkable old lady is Mrs Rachel Garberowitz, who hails from Lithuania. Until the other day she had never been away from her village home, but, at the age of 94, she has at last seen the sea and crossed the Atlantic. At the request of her three married daughters, in Rochester, NY, she came all the way across Europe to Hamburg for Grimsby and stepped on board the *Baltic* at Liverpool on September 4th bound for New York. She was intensely interested in the embarkation of the twelve hundred passengers which the *Baltic* carried and was vastly impressed by the accommodation and fittings of the liner. Mrs Garberowitz is going to live with one of her daughters.[3]

Being a conductress was not a job for the squeamish. The women often secreted money and valuables on their bodies, between their corsets and their underwear, and they resisted changing their clothes during the journey, in case their precious possessions were stolen from them. Everyone hated bad weather; the passengers were unable to get out on deck, away from the omnipresent smell of soup, raw garlic (thought to combat seasickness), unwashed clothes and assorted bodily fluids. Nevertheless, Edith would doggedly eat at least one meal a day in third class, alongside her charges, to ensure the food was palatable.

Conditions in third-class transatlantic ships had continued to improve markedly during the early 1920s, and now bore little resemblance to the horrors of the notorious steerage class before the Great War. On the *Zeeland*, hot seawater baths were available, with special soap that would lather in brine. The women could clean themselves, and wash their clothes. On one occasion an impatient chief steward tried to speed up the process by making two girls share the same bath. Edith insisted each woman should bathe alone, in privacy, and her argument prevailed. Her tussles on behalf of her charges, who were looked down on by some of the crew as racially inferior, often made her unpopular. When she insisted that the third-class women and children were moved to better quarters in the *Zeeland*, to minimise the likelihood of them getting seasick, she encountered hostility from some of the crew, though she won her case: 'Since my women were seeking a new world of hope and freedom, a door to a better life, my thought was that it might just as well start on board ship. The old hands who had dealt with emigrants before 1914 did not exactly approve of my ways: they thought anything would do for "wops" and "bohunks". I did not.'[4]

Edith was sympathetic to the reasons why many of the third-class passengers were making this epic journey to an unknown continent:

We had many Jews – all types – travelling as emigrants from Europe. They looked as if a terror was behind them, running away with a real sense of fear ... all the tragedies of the world seemed to be in their melancholy eyes. They also seemed to have an awful fear of the sea on this, the first time they had ever seen the ocean, or experienced what it could do when in the mood ... How terrible it was for those poor, ground-down peasant types, and the persecuted Jews, to be storm-buffeted on a rolling ship, knowing little of what they might expect, only that it was a land of opportunity that awaited them – a strange land, a better life. Others had gone before, and written home to say so. Difficult to comprehend by those of us who had known nothing but freedom and a comparatively good standard of living.[5]

Edith had other female colleagues on the *Zeeland*, including a Belgian-born matron. She was stout and elderly, with a toothy smile, unshockable and with boundless common sense, Matron had been a licensed prostitute in her youth. She had married well, had grown-up children, and was now a respectable widow who had taken to a working life at sea. Matron and Edith would escort the ship's doctor when he examined the third-class immigrant women. Dr Bayer from Brussels was an urbane character, and the passengers, many of whom had never encountered a doctor before, were reassured to have two female chaperones for their examinations. Health problems abounded in the cramped conditions of third class. Tiny children had often been sewn into their woollen combinations, with just apertures left so they could be held over a chamber pot without being undressed. Edith and Matron would methodically 'unpick' these children in order to bathe them, often discovering skin conditions like impetigo or scabies, which needed medical treatment.

Occasionally women gave birth unexpectedly during the voyage. Heavily pregnant women were not usually allowed to embark, but sometimes their voluminous clothing and deliberate subterfuge would conceal an imminent arrival. Edith relied on the doctor to manage the delivery, but often had to assist. On one trip, Edith encountered a language problem with one very young expectant mother: she was Hungarian, while the doctor only spoke French and Flemish. Edith and the stewardess tried to mime 'push', but met blank bewilderment. Fortunately, a young Hungarian female passenger with a smattering of rudimentary English was located, and translated at the appropriate moment. With urgent instructions being given simultaneously in four languages, the passenger produced a little boy. Babies born mid-ocean were registered by the captain, and their names added to the passenger manifest. If born on a ship sailing under a British flag, the new arrival was registered as a native of Stepney, in London. The arrival of a baby mid-voyage tended to cheer the passengers in all classes and provoke an outbreak of sentimental generosity: one unexpected addition to the passenger list was given an impromptu collection of £450 by the passengers, worth approximately £13,000 today, while another was awarded an unspecified lump sum and a Ford car (worth approximately £3,000 now).[6]

On another transatlantic crossing Edith was summoned to the sick bay where a young woman was evidently on the point of giving birth. No one could locate Dr Bayer – it was cocktail hour and he could have been anywhere – and though the dinner-jacketed purser offered to help, he was rapidly despatched by Edith, who struggled to remove the patient's knee-high boots, revealing a pair of filthy feet. Moments later the baby arrived, and the breathless doctor appeared just in time to cut the cord. The child, a very handsome little boy promptly named Janus, was bathed by Edith. His arrival had taken his mother completely by surprise, so he was wrapped in towels while numerous passengers and crew contributed spare clothing to be cut up and made into a layette for him.

Many of Edith's adult female charges had been recruited in their home countries to be domestics, and were known as Gelley Girls, after the Commissioner of the Canadian Immigration Department who had invented the assisted places scheme. However, some would try to escape their escorts before their intended destination, having arranged clandestinely to meet a boyfriend or a family member. They didn't get far; their clothes made them conspicuously alien, they were unable to speak a word of the host country's language, and they were wearing a ribbon that marked them out as destined for domestic service. The escort system was intended to be protective, so that these young women did not fall into bad company or become illicit 'brothel fodder'.

Edith Sowerbutts was astute, worldly and practical, and dedicated to the welfare of her charges. She was certainly no prude and had spent enough time in and around ports to have a fair grasp of the realities of the sex industry; indeed, she was on friendly terms with the madam of a large brothel in Belgium, while one of her many friends in Antwerp was a former prostitute, now the respectable chatelaine of the ladies' powder room at a smart country club. However, she had an eagle eye for any possibility of sexual exploitation if it threatened her most vulnerable passengers.

On one voyage Edith became suspicious of a man travelling to America with a very young girl who he claimed was his bride. Her documents stated that she was thirteen, but Edith was suspicious as she spent the trip playing with her dolls on the top bunk, and never ventured out of their cabin. On arrival in New York Edith shared her misgivings with the examining nurse from Ellis Island; she too doubted the given age of the 'bride'. Both women suspected that the little girl, very pretty, with long golden hair, had been destined for the sex industry. The nurse reported it to the port authorities, who took action. The man's application to become an American citizen was revoked, but to her frustration Edith never found out what happened to the girl.

Edith also safeguarded unaccompanied children from possible sexual predators on board ships. She would some-times encounter very young girls who were being sent alone to a distant relative in the far country, and who had been placed, with the relevant photograph, on the passport of some unrelated male. The man accompanying them was usually from the same home town or village, and of course this arrangement might be entirely innocent. However, Edith would step in if she discovered that any young girl or boy had been booked into the same cabin as an unrelated adult male. She would move the child to another cabin, to be berthed with a couple of women who spoke their language, and who were willing to take care of them on the voyage.

Picture Brides were another intriguing feature of Edith's shipboard life. These were European-born women who had consented to marry men already living in Canada or the USA, without ever having met them. These women took life-changing decisions after answering a newspaper advert, then exchanging letters and photos, arranging their marriages by post. Edith met one on a voyage to Halifax. Rose was British, pleasant in nature, about thirty-five years old and unmarried;

in the euphemistic phrase of the day, she was 'an unclaimed blessing'. By 1925, Rose saw her chances of matrimony were diminishing, so she replied to a newspaper advertisement placed in a British paper by a widower farmer living in western Canada. She sent her photo in a letter, they corresponded, and agreed to wed. She sailed on the *Zeeland* with a modest trousseau and high hopes. Edith admired her courage, but didn't find out if Rose's future lived up to her dreams.

The Canadian immigration system was well-organised; having interviewed each woman and noted her details, Edith would give her a colour-coded piece of ribbon, which showed her eventual destination: red for Manitoba or Saskatchewan, blue for Ontario, white for the maritime provinces. The women proudly wore these ribbons like badges of honour, or campaign medals, pinned to their clothes. They landed at Pier 21, Halifax, Nova Scotia, to be met by female officers of the immigration department, led by a Mrs Bond, who checked the paperwork provided by Edith and escorted the women and children to trains. They were grouped according to their ribbon colours and then taken to their various destinations all over Canada by so-called 'train girls'. Edith also handed over any unaccompanied children, who had their tickets and vital documents contained in little calico bags securely pinned to their coats.

After leaving Halifax, the *Zeeland* sailed on to New York, and Edith often assisted the American immigration officials on arrival at Ellis Island, although technically her role only covered passengers going to Canada. Ellis Island was a liminal place, where every day thousands of people queued to be processed, and were either cleared for entry and released into the city, or held for deportation. Sometimes would-be immigrants were refused entry on medical grounds, if one member of a family had a communicable disease, such as trachoma, an eye complaint. In those cases the whole family might have to return to the country of origin if they would not separate. This was calamitous; they had usually sold what few assets they owned to scrape together the transatlantic fares, only to be sent back to certain poverty and destitution.

The *Zeeland* expanded its service to take over part of the Irish migrant trade and now the ship ran from Antwerp, calling at Southampton, then Cherbourg, and Queenstown (present-day Cobh) before setting out into the Atlantic. Consequently, Edith had more British and Irish women among her third-class passengers, and she found them more trouble than all the others. Unlike her continental charges, who were examined, bathed, deloused and fumigated before they embarked, the British and Irish women often harboured lice or nits. Edith and Matron would have to treat the lousy promptly, to avoid their 'stowaways' infesting the whole ship.

Health risks were part of the conductress's role, but for most female staff on board, after a cursory medical on joining the line, they could consult the ship's doctor if they were ill. Edith noticed that on every journey, after a certain number of days at sea, all the male crew were summoned in turn to the doctor's cabin for a brief but mysterious medical, and would emerge rebuttoning their trousers. It dawned on her that they were undergoing what was euphemistically known as 'short arms inspection', making sure that they hadn't contracted any sexually transmitted diseases during their last shore leave. She noted the favourite toast of the crew was: 'To our wives and sweethearts ... and may they never meet!'

There was a general wariness about the possibility of sex between women seafarers and their male counterparts, and they were physically segregated within the ship as much as possible. Between the wars the large international ocean-going ships increased the roles undertaken by female staff, although they were still greatly outnumbered by male crew and officers. Female seafarers were expected to be beyond reproach, but, like Caesar's wife, they also had to be *seen* to be blameless. They were not allowed to wear make-up, they must wear a hat or cap while on duty, and they were not allowed to go out with crew members. The women's sleeping quarters, a dead-end corridor lined with twin cabins, were strictly off limits to all men.

Stewardesses generally avoided getting entangled with any male colleague, and if they did have romantic ambitions they would prefer a more advantageous marriage to an officer, or even a passenger. Some women seafarers did find partners afloat, but often the relationship foundered because of the time spent apart, and they separated and returned to sea.

No stewardess would normally risk entering a man's cabin, so male travellers were attended by stewards. Female passengers travelling alone or in pairs were always the primary responsibility of the stewardess; to avoid embarrassment, passengers were advised to ring the bell once if they needed a male steward, and twice for the stewardess. Stewards and stewardesses worked closely together, and often became friends. In married couples' cabins, they would usually share the responsibilities: she would deal with the wife's bed, toiletries, clothes and personal effects, while he cleaned the bathroom and dealt with the husband's belongings. For heterosexual or lesbian stewardesses working on the ships, having a gay male colleague was often an advantage for both parties: they could be friends or allies without any romantic or sexual expectations on either side.

Sexual harassment from men could be a real hazard for

some women. There were those within the ship's company who tried to coerce female staff into having sex, either by offering promotion, or by threatening to sully their all-important reputation if they didn't comply. A number of stewardesses recalled unwanted encounters with questing men. Violet Jessop had to evade the attentions of both an amorous purser and an embittered captain. Edith Sowerbutts was woken one night by a drunken young man who had managed to get into her cabin; 'You're a sailor – I'm a sailor,' he declaimed, as justification. Fortunately, he ambled off, while Edith threw on a dressing gown and went for help. Her 'beau' was promptly detained on deck by two burly crew members, and incarcerated for the night, and Edith moved into a superior cabin, one with a better lock.

There was a general sense of camaraderie among sea-going women in all roles on the passenger ships on the North Atlantic, because they were few in number, and shared common living quarters. Edith had friends of different nationalities throughout her ocean-going career. There was Mrs Nielsen, a tiny, wizened stewardess with flaxen hair, of advanced years, who spoke a variety of Scandinavian languages as well as German. Mrs Nielsen was often to be found at the end of a busy day in her cabin, soaking her aching feet in a bucket of hot water. Another friend was an Irish conductress on White Star Lines, Miss O'Kane, known inevitably as Miss O.'K. One snowy Sunday morning her ship was due to depart from Saint John, New Brunswick. A devout churchgoer, Miss O'K calculated she could attend a service on land before the ship sailed, so she hurriedly threw on some clothes over her pyjamas and set off ashore with just ten Canadian cents for the collection in her purse. Timing was never her strong point; she returned to the dockside to see her ship steaming away, taking all her belongings and documents with it. Forced to borrow money and clothes from the company's agent and his wife, she waited two long weeks until another White Star Line ship put in and she was able to return to Europe.

Edith also had a British-born friend, Emma May Mathieu, a highly competent nurse. She had married a Belgian army officer, but he died in 1925 and she struggled to provide for their two small boys. The children went to boarding school in Brussels while Emma May worked on the ships as a stewardess, though she was impressively over-qualified. She had passed her midwifery final exams, in French, the day after her husband's funeral. Shipping lines often signed up qualified nurses as stewardesses, but denied them 'nursing rank' to save money. Emma May was highly regarded by ships' doctors, so often helped with medical emergencies at sea. It was Edith who persuaded Red Star Line to employ Emma May as a stewardess, a role that brought her ample tips from wealthy passengers, which enabled her to buy an apartment in Antwerp, near the docks.

Emma May had a lively sense of humour; she dealt with one pushy Lothario by coyly inviting him to the wrong cabin, where he burst in on a formidably indignant male passenger, who berated him loudly. On another occasion Emma May and her Liverpool-born sidekick Vera were sitting in a bar in Lisbon, next to a bullring. They were chatted up by a handsome young Portuguese matador in full rig, who was smitten by blonde-haired, blue-eyed Vera. He spoke no English, and Vera's only language was fluent Scouse, but Emma May acted as their interpreter by speaking French. At Vera's prompting, she taught their new friend a couple of English phrases. After a few drinks, Vera and Emma May realised that their ship was about to leave port. Accompanied by their flamboyantly dressed young admirer, they sprinted to the docks, and the two women ran up the gangway with seconds to spare, to the amusement of passengers, crew and onlookers. As the ropes were cast off and the great ship inched away from its mooring, the matador struck a dramatic pose on the sunlit quayside, with his arms outspread, his cape aflutter, and took a deep breath. 'I STICK THE BULL RIGHT UP THE ARSE!' he proclaimed, with a magnificent flourish. This was greeted with a roar of approval and applause from the assembled throng, which the bullfighter acknowledged with an elegant bow, and a proud smile. The chief steward, however, took a dim view, and Vera and Emma May were separated because they were too mischievous to be employed together on the same ship.

Emma May was 'let go' from another ship because she became too friendly with the chief officer. 'It was all that gold braid,' she explained in mitigation. On arrival in Antwerp, as the chief officer's outraged wife stormed up the main gangway to settle the matter, woman to woman, Emma May nipped down the aft gangway with her suitcase, straight into a waiting taxi and headed for Brussels to see her sons. Although usually adept at avoiding trouble, Emma May would occasionally fall victim to sob stories and loan some handsome scoundrel her hard-earned money, never to see him again. As Edith recalled:

She would be the confidante of a duchess, or a friend to a steward or stoker … she had the common touch … she was one of the ship's most valued and respected senior stewardesses. I have rarely met a woman with more 'guts' than she had, and you do meet women of great courage when you go to sea. We started a spontaneous friendship

on that voyage of the old Zeeland in 1926 – it lasted till
1973, when she died, aged 81.[7]

Friendships made with other seafarers were intense, based
on mutual trust and common experiences, but Edith was
aware that it was difficult to sustain any relationship with
land-based friends and family when constantly travelling.
Some of her regular passengers she might see again on other
voyages, but she was frustrated that she never knew what
happened to her thousands of migrant charges, and what
lives they made for themselves in the New World after she
had waved them off.

As for romance, shore-based boyfriends would soon tire
of only seeing her once every thirty days, and female crew
were barred from fraternising with the officers on board.
Edith had a land-based admirer, a Polish Jew who worked in
the company's Antwerp office. He was offered a round trip
to New York, and he accepted because he was planning to
see Edith on board the *Zeeland* on the outward journey.
Unfortunately, he was confined to his cabin with rampant
seasickness, which lasted from the west coast of Ireland till
they reached Nantucket. It was evident that their lives were
not compatible, though, as Edith fondly recalled he had a
gentle old-world courtesy, and was the only man who ever
called her 'a peach'.

One aspect of the seafaring life that especially appealed to
the more adventurous women was the opportunity to explore
foreign cities. After recovering from the voyage and putting
up their tired and aching 'Cunard feet', they would venture
down the gangway to see a show, or go on a shopping spree.
Seafaring women who ate in passenger dining rooms, such as
Edith and Hilda James, had to buy their own evening dresses,

shoes and accessories, as well as uniforms. Edith loved exploring the many cities where they berthed, and was as fascinated by the low life as the high life to be found on shore. In Liverpool she noted the younger male crew members often headed for the city's pubs and dance halls, hell-bent on fun. Senior officers gravely warned the younger crewmen 'to do no more than wave at the "good time" girls on Lime Street', advice that was often ignored.

In Antwerp Edith frequented the opera, but she also liked to visit seamen's bars, always accompanied by some of her shipmates. Most of her time ashore was spent away from the beer and brothel brigade, as she was more friendly with passengers than the roistering crew. However, she was familiar with the red-light district of Antwerp, near the station, which sailors called Ruination Street in a variety of languages. Seated outside the doors, elderly women plied their knitting needles – each one was a madam, touting for business. Edith would pass the time of day with them and they were cordial, knowing she was the conductress, the madamica, from one of the big ships.

New York seemed remarkable to Edith, as it did to many foreign seafarers. It was so different from the dingy hinterland of most British ports in the late 1920s. The streets were brilliantly lit, and lined with giant, colourful advertising hoardings; cinemas and dance halls were emblazoned with lights; art deco skyscrapers loomed over canyons of bustling cabs and automobiles. There was the lure of Hollywood movies, the theatres and music halls of Broadway, the glossiness of the magazines, the hum of commerce, the fashions to be seen on the crowded sidewalks. It was all humming with a sense of industry and optimism, jazz and vibrancy.

Edith frequented speakeasies, informal and unlicensed

drinking clubs, and she would occasionally go dancing at Roseland with her fellow shipmate George. On one occasion Edith was late meeting him, and he guessed correctly that she had joined the throngs of women queuing to see the mortal remains of film star Rudolph Valentino. Her motive was curiosity; Valentino was possibly the most famous film star in the world and he had died suddenly in the Polyclinic Hospital in New York on 23 August 1926. She recalled: 'The movie idol looked shrunken, yellow, wizened … Valentino was 31 when he died, and he looked like any Italian waiter.'[8] She found the lavish grief surrounding his death incomprehensible, but then she maintained no illusions about the man, rather than the matinée idol. Valentino had been a passenger on one of Edith's ships, the *Lapland*, and when they docked at Halifax, he refused to get out of bed to have his papers checked by Canadian immigration authorities as he was an American citizen. Her succinct epitaph was: 'Valentino thought he was immune. He was not.'

Edith often looked after unaccompanied young travellers, and teenagers could be a particular trial. One thirteen-year-old, who was returning to her boarding school in Britain, was entrusted to her care. She was pretty and a good dancer, and Edith as her chaperone had to deter a couple of prowling male passengers. Nevertheless, Edith became suspicious late one evening – she went to the girl's cabin after midnight and found it empty. Two hours later the teenager returned to find Edith waiting for her. She had attended a bachelor's drinks party in his cabin. The man had told the girl to pretend to retire for the night, then to creep along to his cabin when the coast was clear. Fortunately, she hadn't come to any harm, but Edith tackled the 'boyfriend', threatening to tell the captain. The Lothario apologised profusely, muttered something about having young sisters of his own, and kept his distance thereafter.

Edith was occasionally surprised at the behaviour of female passengers when at sea and off the leash, especially when drink was involved. Alcohol inevitably loosened the inhibitions afloat, and during the era of Prohibition in the USA one of the main attractions of travelling on foreign ships was the alcohol freely available once out at sea. One particularly wild sixteen-year-old girl who was travelling alone on the French Line experimented unwisely with the many cocktails available in the bar, and found herself the next day with a cracking hangover and engaged to be married to a Gallic barman whose divorce was pending. Her parents were outraged, and summoned her back to the States in disgrace on Red Star Line, so she was entrusted to Edith's care. She apparently required constant monitoring, and it was a long and acrimonious voyage for the girl, the stewardess and Edith.

By the late 1920s, 'conductress' was not the only role for women seeking sea-going careers beyond the nurturing, caring roles traditionally deemed appropriate for the fairer sex. The boom in transatlantic travel enabled women to take on new roles afloat. An ability to type, coupled with an outgoing personality and organisational abilities, suited women to be stenographers on ocean liners, based in the chief purser's office, where administrative competence was vital.

Stenographers were increasingly employed as a matter of course on many ocean liners, so that passengers could avail themselves of secretarial and administrative assistance while travelling. Edith Sowerbutts credited Canadian Pacific Lines' lady supervisor, Mrs Andrews, with the active recruitment of women stenographers for that company's pursers' offices in 1925. In fact Cunard had been an early pioneer in this field, rather daringly employing a woman stenographer on the *Mauretania* before the Great War, a Miss M. Casey, whose experience subsequently secured her a responsible position working for the Canadian government in Saskatchewan. Similarly, a photograph from 1920 of Chief Purser Spedding on the *Aquitania* shows him seated on deck with his eight male staff. Sitting alongside him is Miss G. Matthews, a young stenographer, wearing a white uniform and stockings, and a cheery smile.[9] For a woman with a portable typewriter and an outgoing personality, working on one of the Atlantic Ferryboats could be a decent way of making a living. By the early 1930s, Cunard actively marketed the on-board skills of its lady stenographers to business travellers, so that they could arrange ad hoc administrative and secretarial support throughout the voyage.

Ships certainly provided floating workplaces for women whose skills were transferable from land. But by the late 1920s there were also some women seeking to take on seafaring roles previously exclusive to men. Although such women were rare – and it is significant that they usually succeeded because they came from privileged and influential backgrounds – they captured the public's imagination and were a frequent source of interest to the press and media.

Victoria Drummond was Britain's first female seagoing marine engineer. She was well-connected, and had her family's backing to pursue her ambitions. The daughter of Captain and the Hon. Mrs Drummond of Angus in Scotland, she was the granddaughter of the First Lord Amherst of Hackney, and a goddaughter of Queen Victoria. Victoria was the first woman in Britain to serve a full apprenticeship, undergoing five years' training at the Caledonian Shipbuilding and Engineering Company in Dundee, under the same conditions as those for boy apprentices, before passing her works test as a journeyman engineer. In 1922 she became the first woman to be admitted to the Institute of Marine Engineers. Despite her commitment, she was failed thirty-one times by the British Board of Trade when sitting her chief engineer exams. Undaunted, she entered the exams run by the Panamanian authorities, as their examiners marked the papers without knowing the candidates' gender and status. This time Victoria passed straight away; she served her apprenticeship and qualified as a ship's engineer in 1924.

Victoria made many long-distance return voyages travelling to South America, Australia and the Far East and, in 1926, she obtained her second engineer's certificate. She was awarded the Lloyd's Bravery Medal and an MBE for heroic actions during the Second World War when her ship the *Bonita* was bombed. Her successful career at sea lasted into the 1950s, despite encountering prejudice and sex discrimination.

Another well-connected young woman sought to take on a maritime career previously exclusive to men. The Hon. Elsie MacKay was a thoroughly modern figure, a marine engineer and a qualified pilot. She was the third daughter of Lord Inchcape, the chairman of the steamship line P&O. 'She is very good-looking and considered one of the best-dressed women and dancers in London society ... one of the most remarkable women of the younger generation, although her modesty has prevented the public from hearing much of her exploits,' enthused the *Manchester Guardian*.[10]

After an early career as a film actress, under the name of Poppy Wyndham, Elsie was appointed to oversee the interior design of twelve of P&O's liners. The vessels were equipped with modern conveniences, such as passenger lifts, electric radiators and air ventilation, but Elsie emphasised the traditional in her interpretation of historic interiors. The *Viceroy of India* was launched in 1929 and in its first-class smoking room Elsie installed oak panelling and a vast fireplace topped with a royal coat of arms, flanked by stained-glass escutcheons in leaded windows, and wrought-iron gates. Apparently, she had drawn on accounts of James I's palace at Bromley-by-Bow, borrowing perhaps from an ironic reference to that monarch's interest in transatlantic travel, coupled with his famous hatred of tobacco. Elsie favoured a variety of historical styles: the music room and the dining room evoked the eighteenth century, while the swimming pool with its classical columns and reliefs recalled the public baths in the recently excavated Italian city of Pompeii.

There was renewed press interest in the possibility of women seafarers taking on the roles of male crew, a consistently newsworthy theme. 'Woman at the Helm' in the *Westminster Gazette*[11] conveyed the revolutionary news that when the Soviet ship *Tovarisch* left Port Talbot docks the previous day, the female third mate, Comrade Diatchenki, was at the helm. Communist Russia was certainly pioneering in this respect: within eighteen months an official communication from Moscow announced that women were to be appointed officers on ships of their mercantile marine. One woman was named as a naval architect in state shipbuilding yards; another was appointed to be captain of a steamer, and was the first woman to command a ship of the Black Sea fleet.[12]

In Britain, as early as 1925, in an article entitled 'Women Sailors', the *Daily Chronicle* reported that a 'woman skipper' had taken her own motor cargo boat down the Thames to the Isle of Wight, and that several foreign ocean-going freighters were known to be commanded by women. In addition, Norway's merchant navy was known to employ women as on-board wireless operators. It was acknowledged that 'Little ships seem to have provided most jobs for women up till now. Scores of Thames barges carry the skipper's wife, a person who is up to any emergency, and as good as most men in the matter of steering, cooking and washing.' The feature condemned the superstition, still retained by some seamen, that women on board brought bad luck to a ship. It also made the salient point that: 'There are plenty of women air pilots, and if the female brain is quick enough to carry out that task successfully, it can navigate a ship with ease, even in time of danger.'[13]

There was much public debate in the 1920s and 1930s about women training to be pilots. Flying as a means of travel was seen as progressive and daring, and women pilots as modern and chic. As early as July 1919, *Cunard Line* reported approvingly on Mrs Leon Errol, wife of the well-known actor, who had created a new record by flying in an Avro aircraft from Hounslow Aerodrome to Southampton in order to embark on the *Aquitania*, as she had missed the boat train. Apparently the ship's passengers gave her an enthusiastic reception when she joined them on board.

The proven ability of women to qualify as private pilots inevitably informed the more trenchant views about whether women could cope with technical jobs at sea. The *Evening Standard* advocated that women should learn to fly, claiming that they were 'better pupils than men ... the cost is only about £20 all told to become a qualified pilot, and that in comparison with other professional trainings is extremely moderate and short, besides being one of the most enjoyable and health-giving careers which any man or woman can take up.'[14] Women who were qualified pilots were also fêted by the press: Miss Lilian Dawson, a fourteen-year-old American girl who sailed to Liverpool on the White Star liner *Megantic*, claimed to be the youngest qualified air pilot in the world. She was a member of the prestigious Pittsburgh Aero Club, alongside the international aviator Charles Lindbergh, and Lilian had obtained her pilot's certificate the previous year. However, by law she was not yet allowed to carry passengers in her plane; as the teenage prodigy remarked, in America a girl was not even allowed a driving licence until she was eighteen.[15]

Flight as a means of passenger transport was already starting to supplant combined train and sea travel across the landmasses of Europe, where 'short hops' between cities were now possible. In 1927 *White Star Magazine* reported on a journey by air from Croydon to Basle and Zurich in one of the big aeroplanes, a twin-engined Handley Page, belonging to Imperial Airways. The writer was impressed by the convenience of being able to get from London to Switzerland within a single day. However, while aeroplanes were gradually becoming a viable means of conveying passengers and their luggage short distances across continental Europe, they could not yet cover the huge expanses of the Atlantic, as it was not possible for them to carry enough fuel for the distance. And while enterprising women might be able to qualify as private pilots, and fly solo across deserts and oceans in small planes, there were still issues about employing women on ships for roles traditionally occupied by men. Various excuses were used; some shipping companies claimed it was not possible to add 'extra facilities' (meaning separate bathrooms) for female technicians, even though they were already provided for those women employed in traditional 'caring' roles such as stewardesses and nurses.

There were other roles afloat that women coveted, but which were denied them. Semaphore and Morse code were taught in the British Girl Guides from 1910, and there was considerable public interest in the relatively new science of telegraphy and wireless operations. Some women sought training as radio officers and were occasionally employed at coastal stations. However, they were not allowed on board British merchant ships, even during the Great War. Although thirty-eight women had passed their radio examinations by October 1917, it was felt they might 'go to pieces' in a crisis.

In 1923 Jessie Kenney, an associate of Emmeline Pankhurst, qualified as a radio officer, having trained at the Rhyl Wireless College. She could not overcome opposition from the Marconi Company, the Board of Trade, or the shipowners to gain work as a radio officer afloat, so she signed on with a shipping company to work as a stewardess, in the hope that she might have the opportunity somehow to prove her skills as a wireless operator once at sea. She found there was no possibility of changing roles, and later recalled that she would often glance wistfully up at the wireless cabin where, in peace and quiet, she could have used the skills she had gained and, in her spare time, continued to study her science in relative peace.

Ironically, competency and increased responsibility were considered in some quarters to be desirable attributes in women afloat in this era. While one transatlantic yachtsman, the maverick Captain Thomas Drake, predicted that 'The day will come when women will command vessels manned by women,'[16] his view remained unorthodox. But some women already working at sea were expanding their knowledge and their skills, in order to become more active crew members. The Board of Trade required every lifeboat or life raft on a liner accommodating forty-one or fewer passengers to have at least two crew members aboard who were qualified to launch it successfully. The awful fate of those aboard the *Titanic*, many of whom perished because of the failure to fill and successfully launch so many of its lifeboats, was still fresh in many people's minds, particularly professional seafarers. In 1929 the training necessary to take charge of a lifeboat was opened to women for the first time. Blanche Tucker was employed as the chief cashier in the French restaurant on board the *Majestic*. She became the first woman to obtain the Board of Trade Lifeboat Certificate. To qualify, as well as proving her theoretical knowledge she undertook a demanding practical examination. Blanche had to prove she could supervise the 'turning out', lowering, handling, sailing and pulling away of a ship's lifeboat containing ten crew members under her command, a considerable display of skill and strength. She recalled:

> I was first asked to describe the contents of a lifeboat, and then to box the compass. The next questions were in relation to sailing a boat, and then I was placed in a lifeboat with a number of seamen and had to take charge of it while it was lowered to the water 70 feet below. Immediately it become water-borne I had to disengage the 'falls' so that it could be got away from the ship's side with expedition. This done, I just took my place with the other members of the crew and pulled an oar. That was the hardest part of the job.[17]

The examiner told one of Blanche's shipmates that he had made the test twice as hard for her as for any male candidates, to avoid allegations of gender favouritism. Blanche did have one advantage: having grown up in the coastal town of

Salcombe, in Devon, she had handled boats from a young age, but successfully launching a large and heavy lifeboat from the side of a liner while commanding ten crew members was a challenge. Nevertheless, she passed and when the *Majestic* next sailed for New York on 7 January 1929, Blanche Tucker was authorised to command Lifeboat 27.

The second seafaring woman to get her 'lifeboat ticket' was a Mrs Berry, a stewardess on the *Olympic*, a company widow whose husband had also worked for White Star. The third was conductress Edith Sowerbutts, who qualified in May 1930. Writing decades later of her decision to undertake this training, resourceful Edith revealed that in her job description, the section headed Lifeboat Drill merely stated that she should 'assist ladies', and gave no further instructions. She thought the best assistance she could offer her charges would be the practical ability to launch and command a lifeboat. Edith was tutored by her officer shipmates, and becoming a certificated boatman was a qualification that gave her great satisfaction.

While Edith relished her role and responsibilities as a transatlantic conductress, by the end of the 1920s she had serious financial worries. Most of her passengers were impoverished would-be émigrées, unaccompanied women and children travelling in third class, and consequently she received very few tips on top of her basic salary. Her income was expended on maintaining a modest but comfortable home for her widowed mother and her rather shy sister Dorothy, and she would join them there between voyages. However, Mrs Sowerbutts had been injured in a road accident, and the family were now struggling to pay the medical and household bills. Drastic measures were called for: Dorothy was winkled out of her dead-end job in an office, which paid a pittance, and sent to sea to make some real money. Edith had approached her old friend Mr Gosling, the victualling super-intendent at Southampton, requesting any available sea-going vacancy for her sister. Such jobs were at a premium; the company's widows had first preference, and applicants needed sea-going connections, but Edith had clout. Sheltered Dorothy was bundled into the Southampton-bound train at Waterloo Station with a small trunk, and embarked on her own maiden voyage, as a stewardess for White Star Line. Perhaps surprisingly, she quickly blossomed into the role; no longer a 'shrinking violet', she proved to be an immense success in her new life.

The ships on which Dorothy sailed mostly offered leisure cruises, linking the east coast of America with the Mediterranean ports, or the West Indies. The stewardess's workload was heavy, with long hours, little free time while afloat, and no paid overtime. However, while the salary was low, a competent and pleasant stewardess could bring home a very respectable income if she earned tips, perhaps as much as £500 or £600 a year. Dorothy Sowerbutts found her new life very congenial; she was pretty, and keen on clothes, and she delighted in the affordable fashions available in the competitively priced downtown stores of New York.

Edith suffered personal tragedies too. In her memoirs she alluded briefly but movingly to her intention to marry a man she had met, but he had passed away unexpectedly, leaving her resigned to be single. She was very depressed about her loss, and wrote: 'in 1929, my spirits were low. The bottom had dropped out of my world. I had just lost a very good friend. My personal sadness, which lasted a long time, had to be very private. I had to carry on with the job regardless.'[18] Edith tried to be philosophical about her loss. 'He died suddenly. Looking back, it would not have worked out. I used to note the fortunate ladies who had so obviously married the men who loved them. I wondered what it might be like, to be so cherished and pampered, mink coats and all. In the end I settled for a modest home of my own, worked for and paid for by all my own efforts.' Nevertheless, Edith was rarely short of male companions, and recalled in her eighties, 'I did not do so badly, with my freckled face and my specs.'[19]

In a career spanning nearly twelve years at sea, Edith took off only two days through seasickness; it was a life that suited her, even though it was physically demanding. Now that the Sowerbutts sisters were both established in decent jobs, they were earning independent salaries and supporting their elderly mother, and, although life was not ideal in many ways, they counted themselves fortunate.

In the case of Edith, it is also apparent that women were gaining new status on board the great ships. Her responsible role as a conductress, the champion and guardian of unaccompanied women and children, occasionally brought her into conflict with some of the crew who looked down on third-class migrants from all over Europe. However, conductresses were the first merchant seawomen to hold officer status, and Edith and her contemporaries often gained respect within the all-male hierarchy on board for their tenacity and determination on behalf of their passengers. Qualifying to 'man' a lifeboat on equal terms with any crew member was also a practical demonstration of Edith's ability and commitment, and gained her grudging respect.

By the end of the 1920s a small number of remarkable and unusual women were choosing maritime careers previously deemed entirely masculine. Those who were successful tended to be well-connected and supported by their families, but nevertheless they were pioneers, demonstrating that women had capabilities and aptitudes largely unsuspected in the halcyon days before the Great War. Victoria Drummond was employed as a fully qualified ship's engineer, and Elsie McKay was both a marine engineer and interior designer on P&O ships. Various new roles for women afloat included occupations that had previously existed only on shore. Now that transatlantic travel was booming, enterprising women, many of them originally from modest backgrounds, went to sea equipped with a serviceable trade or marketable skill. Ann Runcie the hairdresser found she could make a much better living afloat on the Cunard ships than in her native Liverpool, while her fellow Scouser Hilda James was promoted from being a swimming instructress to a cruise directress. By providing valuable services to the female passengers, seafaring women between the wars benefited greatly from the expansion of the transatlantic travel industry.

6

For Leisure and Pleasure

———

Edith admired women travellers of all classes who behaved appropriately and with dignity, no matter what the occasion. Unlike her other female shipmates, the conductress's unique role allowed her to observe the behaviour and pastimes of unaccompanied women passengers in all three classes, and she would flit between the decks several times a day, keeping an eye on her charges. Edith enjoyed dancing, so often joined the *thé dansant*, a popular afternoon activity on ships in the 1920s. Sometimes events could take a startling turn:

> One afternoon the orchestra of the *Arabic* was playing for such an occasion when a sedate, elderly man turned to me and said: 'That girl has just lost her drawers!' There they were, pale mauve silk, at her feet. The elastic must have slipped its moorings. I have seldom seen anyone so cool. With complete nonchalance and considerable dexterity, the young lady retrieved her panties, folded them up and continued dancing with a completely co-operative partner. It so happened she was destined for a minor post in the Foreign Office abroad. Calm, entirely collected, I felt sure she would do well in her chosen career.[1]

Some women's *louche* behaviour irritated Edith. She was particularly disparaging about those who were keen to attract male interest at any cost:

Old girls, young girls and not so young girls; silly old girls, mutton disguised as lamb, all dressed up to the nines, all out on a big safari to ensnare unsuspecting males. They certainly were grist to the mill of hard-up young men – it was a gigolo's paradise at times … I met quite old women, complete with face-lift, tummy-lift and breast-lift, married to nasty younger men, purchased with wealth. A sad sight, those travesties of womanhood. They hung onto their young husbands, kept them well on the leash.[2]

Edith also disapproved of wealthy female passengers fraternising with the men of the ship's company. On some cruise ships the most handsome waiters were occasionally invited by unescorted lady passengers to accompany them on day trips ashore. The women would pick up any expenses, and have a pleasant, attentive young man as company in return. What used to be – and indeed still is – coyly referred to as 'romance' was part of the appeal of travel by liner, and was much vaunted by the shipping lines. Passengers were thrust into physical proximity with strangers for a number of days or even weeks, with ample leisure time to socialise and mingle. They were free of the vigilant observation and possible censure of their own social circles on land, and could reinvent themselves anew, as more charming, witty and better dressed. The environment of the giant ships was subliminally suggestive, with an emphasis on pampering, physical comfort and personal gratification, discreet but attentive service, rich food and wines, the charming dance tunes of the ship's orchestra, and the proximity of the upper boat deck by moonlight, where quiet corners lent themselves to canoodling couples. For some, there may have been an additional frisson to be had from the small but always present risk of physical danger – icebergs, collisions, storms at sea. Where attraction was mutual, both parties were also aware that they only had a finite time afloat, knowledge that often acted as an accelerant to smouldering shipboard romances.

For those hoping for romantic adventure, travel writer Basil Woon recommended embarking at the earliest opportunity to observe one's fellow travellers, and scrutinising the passenger list thoroughly. Despite initial disappointment, he noted that enforced idleness often worked its magic and after a few days of an ocean voyage the most unlikely people would find each other strangely more attractive, probably due to the small pool of potential mates. Once back on dry land, he sagely observed, the magic usually evaporated.

In the 1920s card games were all the rage. A convivial pastime for a small group of people could easily become an avid preoccupation, and for people who enjoyed playing competitive games, such as bridge, a transatlantic trip or ocean cruise was an ideal way to spend one's time. Bridge players would get so involved in the game they would be unaware, or unconcerned, as the ship arrived in some exotic port, only reluctantly leaving the tables to see the delights of Nassau or Havana.

The shipping companies catered for the convenience of serious card players during Prohibition, as many liked to gamble competitively in a comfortable environment, assured of waiter service at the card table. However, one of the hazards of ocean life for keen amateurs was the card-sharp, a professional gambler who made a considerable living by infiltrating gaming tables. Posing as just another passenger, though usually travelling as part of a gang, the card-sharp could be young or old, male or female, elegant or scruffy. They were adept at identifying potential 'marks', usually some exuberant high-rollers at a certain stage of inebriation. If invited to join a game of poker, the professional might feign reluctance or inexperience, and play ineptly, losing small sums of money. They would appear to take their losses philosophically, so their fellow players would gain confidence and commit themselves to bigger and bigger bets. Eventually, through skill, by cheating, or with the assistance of one or more accomplices, the card-sharp would invariably 'scoop the pot'. Seasoned travellers knew to bring their own packs of cards and dice on every voyage, to avoid falling victim to marked packs or weighted dice.

During the early 1920s the *Aquitania* often carried crooks and card-sharps, with at least one gang aboard every voyage, according to Purser Spedding. These characters were well-known to smoking room stewards, pursers and bartenders, despite their disguises. Notices were prominently displayed, warning unsuspecting people to be on their guard. Steamship lines occasionally employed plainclothes detectives to identify those likely to prey on the gullible, while staff would discreetly tip off unwary passengers if they recognised a familiar face or were wise to a ruse. If the captain was alerted that there was a professional card player on board, he had a number of options. Very few of these 'professionals' actually had police records. They were not usually criminals; just extremely good at cards and adept at fooling credulous strangers into thinking they were harmless, until they cleaned them out. The wise captain would invite the suspected card-sharp into his cabin for a friendly chat about inconsequentials, while leaving his large pistol on the desk, unremarked on by both parties. The purser could break up a card game in the public rooms on board ship if bets were being placed, because technically gambling was not allowed there, but it was difficult to police what happened in private staterooms and cabins. Some professional gamblers were tolerated by the steamship lines because they were well-behaved, wouldn't play against naïve youngsters or drunks, were known to play fairly and were liberal with tips to the crew. Others were actively disliked as they could be violent, resistant to interference with their livelihoods. One card-sharp on the *Aquitania* in 1923 was confronted with an accusation of cheating during a game of poker. He instantly attacked his accuser with a broken glass, and nearly cost him his sight. The same character had also attacked Purser Spedding years before on the *Campania*, when he had been exposed as a card-sharp, and had threatened to shoot him.[2]

Edith Sowerbutts recalled one transatlantic trip where one of her charges, the female accomplice of a card-sharp, had an attack of remorse after her partner in crime won a huge sum from a wealthy but inexperienced young man. Just as they docked in New York, the beauty summoned Edith to her cabin and asked her to deliver a thick envelope to the victim's stateroom. Edith deduced that it contained a wad of bank notes, probably enough to get him safely back to London. She speculated that the young woman had felt sorry for the victim, who was probably sadder but wiser for the experience.

There were some female passengers who actively used their charms to their financial advantage, by attracting then compromising some well-off male passenger, and threatening him with blackmail. Basil Woon called them 'sea vamps', and warned susceptible men to be careful not to fall for their wiles: 'This is a profession with the numbers of its adherents swelling yearly. Beware of the beauty travelling alone – be she never so helpless, never so innocent! You'd never believe the number of women who make a regular living travelling back and forth across the Atlantic, preying on passengers of first-class liners.'³ He cited a former chorus girl from New York, who at the age of eighteen hooked a wealthy American businessman into a whirlwind romance in Paris, and accompanied him on the voyage home. While at sea, she discovered he had a wife and family back in Chicago and secretly obtained his address. She had collected a few handwritten *billets doux*, including one compromising note that read 'Honey, I'll be waiting in the smoking-room after dinner. Kisses and Love'. She threatened her victim with exposure to his family, and he paid $10,000 for her silence. It was apparent that a little maritime extortion offered a far more lucrative life than high-kicking in the chorus line, and by 1926 she had completed sixteen round trips on the Atlantic, with an estimated average profit of $1,000 per voyage.

Spedding recalled a similar tale of blackmail, perpetrated by a husband and wife working together. A wealthy French gentleman joined the *Aquitania* at New York, sailing to France in first class. On the second day out, he met an apparently charming married couple. They all had a cocktail together, then the husband made an excuse and left, saying he had to speak to the captain. Gallantly, the Frenchman had another cocktail with the wife; it would have been impolite to leave her on her own, and also she seemed very personable. The trio met before dinner, and again the husband excused himself on some pretext while the Frenchman danced with his wife. After dinner, she boldly suggested they retire to her cabin for some iced champagne, as her husband would stay in the smoking room till the early hours. Inevitably, within minutes the husband burst in, and caught the Frenchman *in flagrante* with the wife. Outraged, the cuckold brandished a revolver, but calmed down at the mention of compensation. The following morning, the chastened victim appealed to the purser for advice. Spedding suggested he cancel the cheque he had already written, but the Frenchman knew that the couple had his home address and were threatening to write to his wife, so he had no choice but to pay for his gullibility.

The many opportunities for young women to advance their interests and possibly improve their financial situation through ocean travel were immortalised in a tongue-in-cheek comic novel, *Gentlemen Prefer Blondes* by Anita Loos, which was first published in the USA in 1925 and became a best-seller. It was written as a spoof journal; the narrator, Lorelei, and her best friend Dorothy are pretty and ambitious flappers living in New York, whose hedonistic lifestyle is funded by presents from a succession of naïve, wealthy men. Lorelei, after an early career in the movies, is now kept afloat financially by Mr Eisman, the Button King of Chicago. The women are not professional courtesans, but rather gifted amateurs, who rely on a judicious mixture of personal charm and playing one 'gentleman' off against the other. Central to the story is their trip on the liner *Majestic*, sailing from New York to visit London and Paris. Mr Eisman, alarmed by a serious rival for Lorelei's affections, has persuaded her to travel to Europe with Dorothy on the grounds that it would be educational, and promises that he will join them in Paris. Lorelei recognises the excellent potential to be found travelling first class on a transatlantic liner. 'I always say that a girl never really looks as well as she does on board a steamship,' and scrutinises the passenger list for new 'gentlemen' whose acquaintance they might like to make.

On board the *Majestic* (which Lorelei admires, because it reminds her of the Ritz, and does not remotely resemble a ship), she is employed by one of her male admirers to charm some confidential military secrets out of another. She manages this adroitly, and presumably to her financial benefit, in the unlikely setting of the boat deck, while Dorothy amuses herself with a tennis champion. After various romantic and remunerative adventures in Paris, London and Vienna, Lorelei makes a conquest of a very wealthy though dull American called Henry Spoffard. Lorelei and Dorothy are whisked back to New York by Mr Eisner, who is concerned both by this new rival, and the bills he is still paying for the girls' European shopping. Lorelei has vowed that she will eschew all male admirers on the return voyage, because she is considering Henry's offer of marriage. However, old habits die hard, and her journal records that she was tipped off about 'a gentleman on the boat who was quite a dealer in unset diamonds from a town called Amsterdam. So I met the gentleman, and we went around together quite a lot, but we had quite a quarrel the night before we landed, so I did not even bother to look at him when I came down the gangplank, and I put the unset diamonds in my handbag so I did not have to declare them at customs.'

Lorelei was not unusual in attempting to smuggle her newly acquired diamonds past customs. American citizens returning to the United States were required to declare all foreign-bought goods over a certain value, and pay duty on them, but many otherwise respectable passengers saw nothing dishonest in evading duty, and they blithely sought the advice of the ship's company. Purser Spedding recalled that women often asked him for help, and he always advised them to declare their purchases, especially in the case of jewellery. There were stern notices all over the *Aquitania* regarding smuggling, and the crew and officers had considerable incentives not to assist in this illegal pursuit. For example, a new diamond necklace bought in Amsterdam for £10,000 would have American duty of £6,000 chargeable on it, bringing the cost to its owner to £16,000. But a passenger might ask a steward or stewardess to help conceal it, and offer them a paltry reward, perhaps a mere £10 or £20. This was a high-risk strategy because, if apprehended, the passenger would be fined three times the value of the goods. In addition, every crew member knew that if they reported the misdemeanour, they would be handsomely rewarded by the Jewellers' Protection Society. One swift wireless message from the steward and the hapless passenger would be apprehended, prosecuted and punitively fined. In the case of the £10,000 diamond necklace, the smuggler would be fined £48,000, and the steward would receive a cheque for £6,400. This was a life-changing sum in an era when a newly-built three-bedroom house in suburban London could be bought for less than £1,000.

So rife was the crime that there were customs and excise agents who travelled the Atlantic in disguise, looking out for smugglers. One woman had bought a great many expensive dresses in Paris, and asked a fellow passenger for his advice on how to avoid paying tax on the gowns, not realising he was a customs officer. He advised her to replace the Parisian labels with Made in New York ones; he even provided her with fake labels. Her new clothes were confiscated at customs, and she was fined $12,000. If she hadn't attempted to deceive and defraud the US government, she could have paid the duty, and kept both the dresses and her good reputation.

Being detained by New York Customs on the basis of a tip-off, especially an anonymous one, could put innocent people to great trouble and inconvenience. It was galling to suffer the indignity of being suspected because some maliciously minded person had given false information to the authorities. One lady, a frequent passenger on the *Aquitania*, was detained while her entire baggage was minutely examined and she was strip-searched. She was not only delayed on the pier for hours but, later on, her home in New York was also searched by revenue officers. Nothing of an incriminating nature was found, and it was believed that all her trouble was caused by false information given by a jealous 'friend'.

While the fictional Lorelei and Dorothy were testing the sexual mores of the era aboard the *Majestic*, in real life the Atlantic Ferry provided transgressive and ambitious women with the opportunity to assert themselves on a foreign shore. The American-born actress Tallulah Bankhead took London by storm in the 1920s. Her reasons for crossing the Atlantic were complex: she was hopelessly in love with an English nobleman, but she was also driven by ambition to succeed on the stage. For a number of years she pushed the boundaries of respectable behaviour, relying on the novelty value of her exotic accent, her physical allure and her larger-than-life personality, and succeeded in both thrilling and scandalising her adopted city, London.

Tallulah started acting on stage in New York in 1918, aged only sixteen. Her father, a US Congressman from Alabama, warned her to avoid men and alcohol; Bankhead later quipped, 'He didn't say anything about women and cocaine.' She quickly gravitated to the Algonquin set, and embarked upon a series of torrid heterosexual and lesbian affairs. She was besotted by an English aristocrat, Napier George Henry Sturt, 3rd Baron Alington, who was studying banking in New York. 'Naps' had served in the RAF in the Great War, rising to the rank of captain. In 1919, on the death of his father, he inherited the title, as well as 18,000 acres in Dorset, but he did not fit the stereotype of the English nobility. Living mostly in his New York apartment, which was known as Naps's Flat to his international coterie of hedonistic friends, he followed a distinctly flamboyant and bohemian lifestyle.

When Naps returned to England, Tallulah was frustrated and restless. She missed him, and though she had been acting on the New York stage for five years, it was a competitive field. A psychic told her that her future lay across the Atlantic; 'Go if you have to swim' was the succinct advice offered. Fortunately, theatrical directors from Europe often visited New York looking for talent, and in 1923 Tallulah was contracted by the impresario Charles Cochran to play the part of a Canadian in a new London play called *The Dancers*. She sailed on the *Majestic*, a vast and opulently appointed liner, which was particularly popular with performers, actresses and musicians from both sides of the Atlantic. There was always a certain frisson among the more sensitive souls about embarking on this particular vessel, as it was the sister ship of the *Titanic* and *Britannic*, both of which had spectacularly come to grief. Perhaps Tallulah Bankhead, who was willing to take life-changing career advice from a self-proclaimed psychic, might have been less sanguine about the voyage if she had known that Violet Jessop, veteran survivor of the sinking of the *Majestic*'s two sister ships, was now working as a first-class stewardess aboard the last of the trio.

Tallulah arrived in London to star opposite Gerald du Maurier, the leading matinée idol on the British stage. His daughter Daphne exclaimed, the first time she encountered Tallulah, 'Daddy, that's the most beautiful girl I've ever seen in my life.' With her glorious hair, her unique voice and accent, her unrestrained dancing and acrobatics, Tallulah quickly conquered the West End. She appeared in more than a dozen plays in London over the next eight years, rapidly acquiring a fan club of 'gallery girls' who mobbed her at the stage door. One of them, Edie Smith, went to work for her as a personal assistant and stayed in her employment for the next three decades.

She starred in the stage version of *The Green Hat* by Michael Arlen, playing a character very like her own persona. The radical heroine, Iris Storm, commits suicide by driving her yellow Hispano-Suiza into a tree at 70 mph, an act of defiance against the two-faced society that has cast her out. The drama critic Hannen Swaffer wrote admiringly: 'She is almost the most modern actress we have.'[4]

'Everything you did was headline news in the 1920s,' observed the presenter Roy Plomley, when interviewing Tallulah in 1964 on BBC Radio's *Desert Island Discs*. While living in Mayfair, Bankhead bought a Bentley, which she drove herself. However, accustomed to the logical grid system of New York streets, she was constantly lost in labyrinthine London. Her solution was to engage a local taxi and pay the cabbie to drive his vehicle to her intended destination, while she tailed him in her own car.

Though Tallulah's relationship with Naps was revived as a result of her moving to London, it was often stormy; on one occasion they met by chance in a nightclub, and he pretended not to know her. (In mitigation, he was escorting his mother; fashionable London nightclubs in the 1920s often attracted clientele of different generations in a way unimaginable nowadays.) Tallulah had her revenge as she swept haughtily past their table: 'So, Lord Alington, you can't recognise me with my clothes on?' she hissed. Their romance eventually foundered, though they remained close friends.

Tallulah's lifestyle epitomised a certain type of celebrity at the time. She was the ultimate Bright Young Thing and consummate party animal, with a passion for bourbon and cocaine, and an unsettling habit of removing all her clothes in public. She smoked four packs of Craven A cigarettes a day, and consequently had a voice that one critic likened to the sound of 'a man pulling his foot out of a bucket of yoghourt'.[5] Idolised by the theatrical world, and a friend of Noël Coward, she was much in demand on the London scene of the 1920s. Off the leash in London, Tallulah was shockingly outspoken but wickedly witty, and she related stories of her latest sexual conquests to the thrilled party-goers of Mayfair. Her alien beauty, consummate acting ability and, above all, the novelty of her accent made her the toast of the town, and allowances were made for her behaviour; had she behaved in the same manner in New York, it is likely she would have been arrested.

But Tallulah's determination to flout conventions, both privately and professionally, did eventually bring her into conflict with the British establishment. In 1926 Tallulah appeared in a stage drama called *Scotch Mist*, playing the promiscuous wife of a British Cabinet minister. The Bishop of London was appalled and complained to the Prime Minister, Stanley Baldwin, but as a result of the scandalised press coverage, the play became a box-office success. Unexpectedly finding herself wealthy, Tallulah acquired the lease on a house in Mayfair – No. 1 Farm Street – and engaged fashionable Syrie Maugham to redesign its interiors. There she threw hedonistic parties that lasted for days, where the participants might be found passed out on the floors or draped languidly over the furniture. Cocaine was the drug of choice for many showbusiness figures between the wars; when asked if it was addictive, Tallulah replied: 'Of course not! I ought to know. I've been using it for years.'

Eventually, serious allegations of 'indecent and unnatural practices' were made against Tallulah, and a report was presented to the Home Secretary in August 1928. The confidential files, which were only released in 2000 by the Public Record Office, reveal that Special Branch detectives searched for incriminating evidence against her on the grounds of public morality. The central allegation was a serious one, that the actress was in the habit of seducing Eton schoolboys on Sunday afternoons, after providing them with cocaine. The rumour was that five or six Eton pupils were accused of 'breaking bounds', being absent without permission from school after being picked up by car, in order to meet the actress at the nearby Hotel de Paris in Bray. While the details are confused, there may have been some truth to this story as other sources claim that the Eton authorities objected to Tallulah providing cocaine to the schoolboys before chapel, as it made them ungovernable during evensong. MI5 had received a copy of a circumspect private letter written by the headmaster of Eton to a number of parents, denying that any boys had recently been expelled, but admitting that two boys had been 'dismissed' and a further three 'disciplined' on the unlikely grounds of having infringed the school rules about motoring.

But once subject to outside scrutiny, the British establishment quickly closed ranks; the investigators found no witnesses or evidence at either the hotel or the school, and they eventually reported: 'No information could be obtained at Eton ... the headmaster is obviously not prepared to assist the Home Office – he wants to do everything possible to keep Eton out of the scandal.' They concluded that Tallulah

Bankhead was 'an extremely immoral woman', but the investigation seems to have gone no further. However, Special Branch missed, or perhaps ignored, an odd and possibly significant coincidence in this story. The highly respectable headmaster of Eton, who had resolutely refused to co-operate with the investigators, thereby stopping the enquiry in its tracks, was Dr Cyril Alington. He was descended from a long line of unimpeachable clerics, but he was also related to Tallulah's former lover, Naps, the 3rd Baron Alington. In November 1928, three months after Special Branch failed in their attempts to have Tallulah thrown out of Britain as a danger to public morality, Naps married Lady Mary Sibell Ashley-Cooper, the daughter of the 9th Earl of Shaftesbury.

Tallulah's reputation for outrageous behaviour was well-founded, and she walked a fine line between being admired for her chutzpah and damned for her conduct. She had been fêted and rewarded in London, but with Naps now married to the daughter of an earl, and the father of a daughter born in 1929, Tallulah realised she had finally run out of road in Britain. She was offered a lucrative Hollywood contract of $5,000 a week by Paramount Pictures, and in January 1931 she sailed back to America. Her sojourn in Britain had brought her public notoriety, largely based on her novelty and 'otherness', as well as considerable stage success.

For Josephine Baker, crossing the Atlantic completely transformed her life. As an African-American woman born in poverty in 1906 in St Louis, she had a difficult childhood. She was sent to be a housemaid aged only eight years old, but was badly treated by her employer. She witnessed the horrific violence of the St Louis race riots, and was briefly married at thirteen, then ran away to Harlem in New York, where she started dancing for nickels to entertain queues of people waiting to get into music halls. Entirely self-taught, she turned professional aged fourteen. In New York, aged nineteen, while appearing at the Plantation Club, Josephine was offered a place in a new dance show produced by an American called Caroline Dudley, a wealthy white socialite and frequent visitor to Harlem, who was recruiting black performers and musicians for a show featuring jazz music and dance. The show, *La Revue Nègre*, was to be staged in Paris. Mainstream France and Great Britain in the 1920s and 1930s had an ambivalent attitude to black culture, but Paris was generally much more welcoming than the major US cities, where discrimination was part of the everyday experience. African-American jazz musicians were very popular in the entertainment world of Paris by the mid-1920s, and profession al performers flocked to the *laissez-faire* cosmopolitan nightclubs and music halls of the most sophisticated city in the world. The exuberant style of music and dance embodied the great sense of relief at the end of the war, and the artistes were admired for their creativity and skills rather than judged by the colour of their skin.

Josephine Baker sailed for France on the *Berengaria* on 15 September 1925, along with twenty-four black musicians. The *Berengaria* was previously the German flagship *Imperator*, but had been handed over to Cunard after the Great War as part of the reparations agreement for a discounted fee of £500,000, and was now the largest of the line. The previous year, Cunard had recognised that there was a burgeoning market in American tourist-class passengers from the States, travellers who required comfortable but affordable voyages to Europe. Cunard therefore upgraded the *Berengaria*'s third-class facilities, providing better victualling for all passengers, refitted cabins and enhanced waiter service in the restaurant.

The Cunard company minuted the popularity of this new class of travel for Americans heading east on a restricted budget, and its affordability enabled performers such as Josephine and her fellow musicians, as well as students and teachers, to travel across the Atlantic in relative comfort, and in increasing numbers.

The *Berengaria* arrived in France after seven days afloat, and rehearsals for the show started immediately. Josephine Baker's extraordinary style of dancing – a blend of sinuous hip-grinding and energetic gyrating, cakewalking and tap dance – quickly made her a star. Her inaugural performance was on 2 October 1925 at the Théâtre des Champs-Élysées. She performed a *danse sauvage*, bare-breasted and wearing little more than pearls and feathers, to an initially stunned and then rapturous Parisian audience. Josephine Baker was an overnight sensation; within two years of her arrival in Paris she was the acclaimed star of Paris's legendary cabaret hall, the Folies Bergère, where she became both celebrated and notorious for her 'Banana Dance' (performed topless, with a 'skirt' of bananas). The French admired her chic appearance, her phenomenal stamina, her inventiveness and humour. Picasso and Hemingway were fans; she was known in the press as the Bronze Venus and admired as an accomplished artiste, a star and a socialite. While some black contemporaries criticised her for perpetuating racial stereotypes, Josephine knowingly used the tropes and imagery of the day in a way that ridiculed prejudices and knocked down the barriers of segregation. Her success on stage allowed her to open her own nightclub in Montmartre, Chez Josephine, which was small, exclusive and very expensive.

Now a national celebrity in her adopted country, Josephine accepted lucrative sponsorship deals to endorse beauty products. During her time in France she learned French, Italian and Russian, and how to fly her own two-seater plane and she starred in four movies. Tellingly, she never made a Hollywood film, but in France she became a cultural icon, fêted in a way that could never have happened at that time in the country of her birth. Indeed, it was on a transatlantic voyage back to America in the mid-1930s, where she had been lured by a lucrative offer to appear in the Ziegfeld Follies, that Josephine was reminded of the racial discrimination then inherent in her own country. An unnamed movie actress travelling on the same ship refused to dine at the same table as Josephine Baker, on the grounds of her colour. After fulfilling her professional obligations in America, Josephine Baker was glad to return to France, where her stellar talent was more important than her race, and her achievements brought her acclaim, wealth and instant recognition.

By the 1920s the nature of fame itself was changing. With the arrival of the mass media, illustrated newspapers and newsreel films, a person could become a celebrity almost overnight. In previous eras achieving fame was a slow, almost sedimentary process; people became renowned for what they did, such as winning wars, writing epic poetry or inventing a better mouse-trap. By the 1920s and 1930s, because of the growth of photography, film and popular culture, being famous was a matter of being instantly recognisable, often all over the world. Charlie Chaplin, shortly after making his first silent films in Hollywood, visited New York and was amazed to see his image on giant billboards and posters all over town. In real life he was unrecognisable without his make-up, but his alter ego was everywhere, instantly identifiable by millions of enraptured strangers who were unaware of his very existence just weeks before.

One of the many attractions of transatlantic ocean travel for a certain sort of person was the possibility of rubbing shoulders with the famous. First-class travellers avidly scanned the printed passenger list, a copy of which was supplied to each cabin. Along with those who were household names only to their own relatives, they might find an intoxicating mix of royalties, aristocrats, heads of state, politicians, noted beauties, captains of industry, sporting heroes, millionaires and maharajahs, and stars of the stage and silver screen, all aboard on the same voyage. Second-class passengers' names were also listed and provided to first-class cabins; there was a certain amount of social permeability between the two superior strata of accommodation. However, no reference was made to the vast majority travelling below in third class; it was assumed that they would be of no interest to their social superiors.

The choice of ship was important too. Smaller vessels tended to be more friendly in atmosphere, and passengers with large but fragile egos often preferred to lord it on more intimate ships, rather than risk being out-gunned by more stellar personalities on the bigger liners. The larger ships had a reputation for attracting cliques, especially in first class. Common interests quickly led to the formation of social groups; there were the sports enthusiasts, bridge players, poker fans, the steady drinkers, the dance fanatics, the seasickness sufferers and the inveterate gossips. 'The Atlantic is rich in personalities. Most of the great of the earth at some time or another have been seasick on its bosom,'[6] remarked Basil Woon. Famous authors, thespians and performers often attracted a coterie of fellow passengers keen to scrape acquaintance. However, star-spotting on an ocean liner was not always plain sailing. While the public areas and deck spaces were designed so that people could mingle, some VIPs preferred to travel in cloistered privacy. Large self-contained suites with individual balconies and high-end, discreet service allowed the reclusive passenger to avoid public scrutiny for the entire voyage, if desired. Enigmatic actress Greta Garbo rarely left her cabin while afloat, and on one occasion managed to avoid the waiting press altogether by disembarking in a borrowed stewardess's uniform. Film stars needed to be beautifully dressed and coiffed when 'on duty'; Marlene Dietrich only appeared in public at dinner, though she did time her arrival for maximum effect, taking to heart the advice of her friend Noël Coward: 'Always be seen, dear, always be seen.'

The better-known liners cultivated their pet celebrities, offering them preferential rates. The fierce commercial tussle between various shipping lines to convey eight glamorous Ziegfeld Follies girls to Europe in 1923 was won by French Line, whose triumphant marketing declared 'every French boat is a little Paris'. European royalty held an exotic allure for many; Queen Marie of Romania was a frequent Atlantic traveller who was willing and able to 'play to the gallery'. Hints about their aristocratic fellow passengers were used to attract aspirational potential customers. The shipping companies' PR departments made the most of well-known personalities, taking photos of their glamorous shipboard life, and issuing images and interviews to the press. Those who had no intention of meeting their fans would travel with a posse of pals to keep interlopers at bay. Disappointed observers could nevertheless closely watch their habits when they did appear in public. One might note that the Prince of Wales had adopted the American fashion of only eating with a fork held in his right hand, that Charlie Chaplin favoured heavyweight reading matter such as Stoddard's *Revolt Against Civilisation*, or that Gloria Swanson's husband, the Marquis Henri de la Falaise, was an accomplished tango dancer. Such small but authentic details, providing proof of proximity to celebrity, were often the cherished highlights of the voyage and would be retailed to their friends and family on the passengers' return.

Familiar faces often patronised particular shipping lines because they liked being attended by the same stewards, stewardesses and pursers. Gloria Swanson favoured the chic vessels of the French Line. Percy Rockefeller liked the *Olympic*. The *France* was a particular favourite with actors, writers, singers and society types, but business travellers tended to prefer the massive Cunard ships such as the *Aquitania* or *Mauretania*, or American ships such as the *George Washington*. Anna Pavlova favoured the *Leviathan*. The Prince of Wales liked the *Berengaria*, while his mistress, Thelma Lady Furness, preferred the *Majestic*. Brigadier-General and Mrs Cornelius Vanderbilt, the acknowledged leaders of New York society, patronised the *Mauretania*.

Cunard was the first shipping line to introduce its own on-board photographers in the late 1920s. Casimir Watkins came up with the idea while sharing a mid-Atlantic cocktail with a Cunard director on the *Berengaria*. Borrowing £500 from an uncle, he set up a photographic company called Ocean Pictures. Their first studio was on the *Lancastria*, but they later acquired the exclusive rights to provide professional on-board photographic services on the prestigious *Queen Mary* and the *Queen Elizabeth*. The company's professional snappers were briefed to take attractive pictures of the passengers on these prestigious vessels, but they had to behave with decorum and ensure the images flattered the subject. The photographers would work through the night to develop the negatives and make multiple prints in the ship's darkrooms, so that passengers could order copies for themselves the following morning for a modest fee. The best shots would then be released to the press when the ship docked; a photo-opportunity was the *quid pro quo* for celebrities, and usually they were happy to co-operate with the company's press initiatives, seeing it as part of their own publicity. For ordinary passengers the presence of professional photographers on the ship added to the glamour of the experience. The 1920s were a golden age of photography, when people wanted affordable, flattering portraits of themselves, as mementoes of a particular voyage, and as gifts for friends and family.

The docking of any prestigious liner always attracted a mob of newsmen and photographers, known as the Gangplank Willies, desperate to get on board to interview celebrities of all types and nationalities. The first-class section of the ship would be besieged by journalists and photographers, swarming on to the ship like pirates seeking their bounty. Hard cash would be offered to hitch a ride on any boat going out to meet an incoming liner, in order to be the first aboard. The New York papers had dedicated news desks covering the arrivals and sailings of the famous on international ships. The journalists' speciality was to startle a celebrity into providing a revealing quote or piercing *aperçu* – as one news-paperman remarked, 'An ocean trip makes people want to talk.'

Shipping companies deliberately cultivated the press in order to garner valuable publicity for their vessels, and the calibre of stars and VIPs to be found on them. With the growth of portable cameras and flashbulbs, photographers were able to take shots of celebrities as they disembarked in

Southampton, London or New York, and reporters from the papers could glean a few words from a notable if they met them at the gangway. There was often an unseemly scramble for pole position on the dockside when it was known a particularly photogenic or newsworthy individual was setting foot on dry land, but it was a symbiotic relationship as both the celebrity and the media outlet benefited from the encounter. As Cunard's Commodore Bisset remarked, 'Very few celebrities are shy of reporters. That is one reason why they are celebrities.'[7] Evelyn Waugh wrote:

> The classic ground for the sport is a liner arriving in New York. New Yorkers still retain a friendly curiosity about their foreign visitors – indeed, believe it or not, a bulletin is printed and daily pushed under your door in the chief hotels, telling you just what celebrities are in town, where they are staying, and nominating a Celebrity of the Day.[8]
>
> To satisfy this human appetite, the reporters come on board with the first officials and have ample time before the ship finally berths to prosecute their quest. They are not got up to please. Indeed, their appearance is … a stark reminder of real life after five days during which one has seen no one who was not either elegantly dressed or neatly uniformed. American papers have at their command most prepossessing creatures of both sexes, but they choose only those who look like murderers to greet visitors. They are elderly and, one supposes, embittered men. They have not advanced far in their profession, and their business is exclusively with the successful. Their revenge is a ruthless professionalism. They look the passengers over, and make their choice, like fish-brokers at a market. One of their number, the grimmest, stalks into the lounge, breaks into

a distinguished group, taps an ambassador on the arm and says, 'The boys want a word with you outside.'[9]

Of course, the vast majority of transatlantic passengers were not celebrities, and their reasons for travel, no matter how pressing and potentially life-changing to them as individuals, held no interest for the Gangplank Willies, whose natural prey were already household names.

However, transatlantic travel between the wars did provide the means and the opportunity for ambitious but impecunious American women looking for opportunities abroad, and Paris drew them like a magnet. The irrepressible collector of international celebrities, Elsa Maxwell, made a lucrative career as a professional party planner, gossip columnist, press agent and impresario. Her success lay in avidly cultivating the wealthy and introducing them to rising stars, thereby improving her patrons' social standing and assisting her protégés' finances. Bumptious, homely looking and publicity-mad, her living came from 'gifts' and 'loans' from the rich. For decades Elsa frequently travelled the Atlantic with her British-born lover, the socialite, heiress and classical singer Dorothy Fellowes-Gordon, known as 'Dickie', forging alliances between the wealthy but socially gauche, and the more decorative but cash-strapped figures of the world of showbusiness.

Paris was the scene of her first great success: shortly after the Armistice, Elsa, who was living in New York with her girlfriend, was asked by Mrs Edward Stotesbury, the American banker's wife, to accompany her recently divorced daughter, Mrs Louise Brooks, across the Atlantic to Paris. Louise needed relaunching in European society, as there was considerable stigma attached to divorce in the United States. The Stotesburys bankrolled the trip, paying for Elsa's passage, and they all stayed at their family house in the Rue des Saints-Pères. The scintillating parties that Elsa threw on behalf of her client in Paris established the young woman at the heart of the city's social life. Louise fell in love with Douglas MacArthur, the youngest brigadier-general in the US Army, and they married in 1922.

Elsa was suddenly in demand everywhere in society, and many wealthy people wanted her to arrange parties of all sorts for them. They imagined that this fat, jolly, bossy woman was important and well-connected. By pretending to be indispensable, Elsa made herself so. She was essentially a self-made woman, a social entrepreneur from a nebulous background, who created a lucrative role for herself as a 'fixer' and party planner. Her invention of the novelty party was adopted wholeheartedly by the Bright Young Things of the 1920s. Treasure hunts organised by Elsa guaranteed the participation of the gilded young, hell-bent on fun, competing to locate the pompom off a sailor's hat, the hairs from an admiral's moustache, or the undergarments of a fashionable *soubrette*. She also created the murder mystery party, a thoroughly British phenomenon that took London society by storm, as it combined snobbery with violence. Her events provided thousands of words of copy for newspaper gossip columnists, helpfully tipped off in advance by Elsa herself, and their printed stories titillated or scandalised the readers.

Elsa found notoriety through novelties such as the 'Come As You Were' party in Paris, in 1927. Elsa's messengers handed an invitation to each of sixty guests at random hours of the day and night. Each recipient was asked to attend dressed exactly as they were when they received their invitation. The Marquis de Polignac cut a dashing figure in full evening attire, except for his missing trousers. Daisy Fellowes had her lace pants in her hand. Bébé Bérard wore a dressing gown, had a telephone attached to his ear, and shaving cream on his face. Several gentlemen who rated honour above vanity attended in hairnets. The party was the sensation of Paris.

It was rumoured that Elsa Maxwell always crossed the Atlantic with fourteen trunks and a hatbox – the trunks for her press clippings, and the hatbox for her other dress. It was true that she had no interest in clothes and always wore either business attire of matching skirts and jackets, or a $20 evening dress picked off the rack of a department store. The reason was simple: her clients and friends were inevitably chic, slim and well-dressed, while she was burly and rather plain. By always wearing the same simple clothes, Elsa did not compete; instead she cultivated a deliberate impression of competency and continuity. However, she did harbour a rakish taste for cross-dressing, and rarely resisted the opportunity to appear as some particularly masculine figure from history, such as Napoleon, at one of the many fancy-dress parties she pioneered. Appropriately enough, fancy-dress events became regular features of shipboard entertainments throughout the 1920s and 1930s.

So famous did Elsa Maxwell become as a publicist and 'fixer' that she was offered an unusual commission to liven up the sleepy Italian resort of the Lido of Venice. This sandy strip of large hotels and beach huts, located a short boat-ride from the ancient maritime city of Venice, had long been a summer resort for the Italian middle classes, providing families with a relaxing seaside break, interspersed with occasional cultural excursions across the lagoon to the treasure trove of Venice itself. Elsa accepted the challenge; she greatly valued what she called 'the hypodermic value of an occasional celebrity', and was also adept at spotting talent among her showbusiness acquaintances. By calling in some favours from her elite clientele, she threw a very well-attended party in the Venice Lido for the fashionable and popular Queen Marie of Romania in 1921, at the stylish Hotel Excelsior. The following summer, 1922, Elsa further developed the resort's trend-setting reputation by returning with Dickie, and 'a young man with an unusual, almost Mongolian countenance', whom they had met at a party given in Oxford by Sibyl Colefax. The young man was Noël Coward, and he accompanied them as their guest to the Lido where Elsa and Dickie had been commissioned to organise a society party for the Duke of Spoleto. The party was a great success, and thanks to Elsa's efforts, Venice and the Lido became the fashionable place for the international set to gather in the summer, the romantic haunt of Cole Porter, Emerald Cunard and Lady Diana Cooper.

By constantly flitting between America and Europe, Elsa made herself the doyenne of international society and showbiz. In later years she based herself at the Waldorf Astoria Hotel in New York, but she was equally at home in Paris, mingling with the 'beautiful people'. While aspirational Europeans of all stripes were heading to America hoping to be 'discovered', there was a corresponding wave of Americans heading for Europe, often prompted by the much reported activities of social fixers such as Elsa Maxwell and her coterie of the international set. Europe as a fascinating, vibrant destination increasingly appealed to Americans of all ranks and incomes in the mid-1920s.

Writer Basil Woon noted in 1926: 'Drinks, divorces and dresses are the principal reasons why Americans go to Europe.'[10] Transatlantic travel to the Old World attracted the hedonistic, the curious and the newly single. In America there was a renewed general interest in overseas travel. 'How're you gonna keep them down on the farm, Now that they've seen Paree?' was a popular vaudeville song in the States, and reflected the desire of ordinary Americans to explore Europe, often – though not necessarily – for the most high-minded reasons. Cities like Paris, London, Berlin and Amsterdam offered ample opportunities for cultural tourism as well as leisure and pleasure, in exchange for hard foreign currency, and a few American dollars went a long way in post-war Europe. The exuberant counter-culture of showbusiness, dance, music and especially avant-garde art particularly attracted Americans to Paris. They read tantalising articles in their newspapers and magazines about the artistic avant-garde, whose aspiring painters and writers could live on a few dollars a day in the more bohemian quarters of Paris, the world's most beautiful city. There was wonderful cuisine available for a fraction of what it would cost at home, and an artist could be free of the strictures of Prohibition.

In addition, there was the lure of the progressive music scene: African-American musicians were welcomed for their abilities, and jazz, that heady blend of focused discipline and inspired serendipity, was all the rage. It was not surprising that the American composer George Gershwin created 'An American in Paris', a jazz-influenced orchestral piece of great verve, whose inaugural performance in 1928 included four Parisian taxi horns. Paris was the magnet for the international wealthy and fashionable elite too in the 1920s and 1930s. During the war years they had been cut off from their beloved Paris. With the return of peace they flocked to the city to buy clothes and accessories, indulging themselves in the fashion houses of Worth, Lanvin and Poiret, their natural habitat. But clothes were also an important aspect of the culture on board the ocean-going ships, as Emilie Grigsby, a Kentucky-born heiress, fashion maven and active international socialite, firmly believed. Emilie frequently sailed on the *Olympic* and the *Aquitania* in the 1920s, between her home in New York and Cherbourg, the disembarkation port for Paris, in order to visit the greatest French couturiers such as Jeanne Lanvin and Madeleine Vionnet. Her taste in clothes was daringly advanced for the time. Emilie was petite, and very elegant; she favoured full-length evening gowns in iridescent colours, which gave her a regal appearance. Renowned for her pale-skinned beauty and reddish-golden hair, she instigated the trend for making spectacular entrances on board ship, using the sumptuous surroundings almost as a stage setting to show off her wonderful clothes and jewellery.

Essential to the fashionable woman on board an ocean liner was the opportunity to make a stylish, almost dramatic entrance, and ships' architects created set-piece staircases and mirrored lobbies where exquisitely dressed passengers could pose in their finery. The *grande descente* was the dramatic staircase leading down to the public first-class spaces of large liners, particularly the saloon, restaurant or ballroom. It acted like a catwalk, enabling the well-dressed to make a glamorous entrance every evening. Cutting a fashionable figure in front of one's fellow passengers became part of the experience of high-end transatlantic travel, and so synonymous were fashion and liners that by the 1930s some couturiers chose to stage runway shows with models on board the big ships.

By the mid-1920s many independent professional women crossed the Atlantic in order to enhance their careers, particularly in the fields of the fashion and textiles industries. British and American businesswomen especially benefited from the frequency, ease and convenience of the Atlantic

Ferry, and the common language was an advantage for those seeking new areas of commerce and trade. Frequent transatlantic travellers to Europe included American buyers, the merchandise selectors engaged by the large department stores in major cities in the USA. They were employed to source and negotiate to buy luxury European goods, principally high-end fashions for men and women, and their buying power was considerable. These overseas buyers descended on London and Paris in their hundreds several times a year, often travelling together to attend trade fairs and fashion shows. There was a huge market for French fashions and English tailoring among the American elite: both France and Britain were seen as chic in the US, and the strength of the dollar made luxury goods saleable in the American market, even after the imposition of import taxes. France was considered the best source of clothes for women and children, and related products such as millinery, accessories, lingerie, shoes, jewellery, tableware, upholstery fabrics, household linens and indeed textiles of all sorts were in high demand. Men's wear for the fashionable American tended to be a combination of French and English styles, tailoring fabrics, tweeds, flannels, neckties.

About one-third of fashion and textiles buyers making these regular transatlantic trips were women, travelling on behalf of American department stores, or for their own retail businesses. They tended to be well-paid, resourceful and hardworking. As it was a competitive business, each buyer jealously guarded her industry contacts. Fashion cycles meant that buyers picked the styles they wanted for future seasons, three, six, nine or twelve months ahead of delivery. They would view a designer's collection, select those items ideal for their clients, and place their orders accordingly. Buyers could request particular modifications to suit American tastes, and they needed confidence and a thorough knowledge of their own market before committing thousands of dollars to buying future stock. Making the wrong choice about colours, silhouettes or fabrics could be financially disastrous, but getting the right collection was immensely profitable. The buyer's role in acquiring French and English fashions was highly influential on the development of the American fashion and textiles market, as the cut-price garment industry was quick to copy and mass-produce more affordable versions of 'Paris modes'.

American fashion and textiles buyers would descend on London and Paris in two particular seasons, usually January, and late July to early August. Making at least two buying trips a year made certain transatlantic ships a sort of floating business hotel – a place to sleep, eat, read, gossip, play cards and plan one's buying strategy. Card games included pinochle and poker, and buyers often liked taking part in the pool, the betting game based on guessing the distance covered by the ship every day. They tended to be sociable and liked congregating in the smoking rooms. Buyers were popular with the crew – they were realistic in their expectations, often coming themselves from more humble backgrounds, but appreciative of good service and consequently good tippers.

Interestingly, many of the women crossing the Atlantic on buying trips were not in the first flush of youth: experience and commercial ability counted for far more than youthful charm. One veteran American traveller was Mrs Mary Jane McShane, who was lauded in *White Star Magazine* in October 1928. She was seventy-four years old, and had crossed from New York to Southampton on the *Olympic* the previous month. Described as an 'energetic old lady', this was Mrs

McShane's fiftieth voyage on White Star, and she was on the perennial hunt for British stock for her antiques shop, which she had been running for fifty-four years.

Transatlantic travel also enabled British businesswomen to expand their professional horizons. London society hostess Sibyl Colefax first sailed to America in November 1926 at the age of fifty-two. Her primary aim was to visit her son, Peter, who was working in New York. However, she was also nurturing ambitions to set up an interior design business that she could run profitably, as her husband Arthur, a London barrister, was growing deaf and struggling to earn enough to fund her ambitious programme of entertaining. Months earlier, while visiting Paris, Sibyl had met Elsie de Wolfe, a highly successful decorator and interior designer in her sixties. Elsie had 'a shop full of beautiful things', and an exclusive and impressive client list, many of whom were already part of Sibyl's social circle. Sibyl also had excellent taste, and had created an exquisite family home out of a miniature eighteenth-century manor house on the King's Road, in Chelsea. It was full of antiques, subtle textiles, venerable ceramics, lacquered Oriental furniture, reclaimed panelling and a historic staircase, with subtle but effective modern lighting, all achieved on a restricted budget. Following her renewed acquaintance with Elsie in New York, Sibyl planned her own future as an interior designer.

Sibyl lunched with her friend Noël Coward, visited the opera and theatre, and spent Thanksgiving with the Cole Porters. Taking the train to the west coast, she made influential new friends in the Hollywood film industry, including Gloria Swanson, Mary Pickford, Douglas Fairbanks Jr and British-born Charlie Chaplin. Her return to the east coast brought her into contact with some extremely wealthy

American owners of Old Master paintings (those executed before 1800), through the auspices of the London-based art dealer Joseph Duveen. Sibyl had an excellent visual memory, and on her next trip to New York, in 1928, she was asked by the Royal Academy in London if she would act as their go-between, approaching American collectors to persuade them to lend their precious paintings. She had met a number of influential museum curators and private collectors on her previous visit, and her knowledge of their holdings was invaluable to the curators of the exhibition on great Italian masters, which was held at the Royal Academy in 1929. As a result of her two successful transatlantic trips, Sibyl gained confidence in her commercial knowledge and aesthetic judgement. She resolved to forge a new career in the American manner, preparing the ground and using her newly-acquired contacts in the antiques and fine arts trade. She opened her London business, Sibyl Colefax Ltd, and quickly established herself as a professional interior decorator, catering for the wealthiest clients in her social milieu.

Women's experience of transatlantic travel in the mid-1920s is often depicted as fun and frivolity, an impression often reinforced by popular writers of the time, such as Anita Loos and P.G. Wodehouse. Certainly, the publicity images commissioned by many of the major shipping companies at this time depict the pleasure and enjoyment to be had from the experience. Simultaneously, the growth of photography and the new media invested the business of transatlantic travel with a scintillating element of glamour, thanks to the co-operation and collusion of international stars from stage and screen, who travelled on the great ships as a necessary element of their careers. Celebrities and the achingly fashionable naturally saw their shipboard appearances as just another aspect of their lives lived partly in the admiring gaze of the public. But the ships also provided the means by which ambitious, restless women could slip the bonds of their old lives and strike out towards an independent future. Tallulah Bankhead and Josephine Baker crossed the ocean in order to reinvent themselves professionally and personally in other countries, using their talents and their unique abilities to achieve public acclaim. Elsa Maxwell closely observed and exploited the restless, migratory habits of the international wealthy elite, and established herself as their quintessential social fixer on two continents. Sibyl Colefax learned by example from encountering dynamic, pioneering business women in America, and returned to Britain to launch an interior design company in her late fifties. As the 1920s drew to a close, the booming business of frequent, fast and reliable ocean travel between Europe and North America, and the many businesswomen who benefited from the phenomenon culturally, financially and personally, changed the life chances of subsequent generations of women.

7

Depression and Determination

———

The Great Depression coloured every aspect of the early 1930s, spreading rapidly through the industrialised nations of the west like a darkening stain. Throughout the North American continent and Europe, banks failed, companies folded, factories laid off staff overnight and great shipping companies mothballed half-built vessels. As unemployment soared, the financial crisis brought massive disruption to international trade and commerce.

The Depression had serious repercussions for transatlantic travel. Experienced working women who relied on the great ships for their livelihoods, such as stewardesses, were made redundant by the recession, or had to accept less favourable posts at sea in order to maintain their households. Female passengers in third class – particularly the young and adaptable who had grown up in poor communities but already had a foothold in America through a relative already established there – had been propelled to emigrate by the worsening prospects in their own countries, in the hope of improving their life chances. The financial situation galvanised the professional decisions of women with broader career aspirations, such as writers, performers and entrepreneurs, who needed to travel the world for literary tours, personal appearances and other engagements. For them, the Atlantic Ferry was the means to establish useful contacts, to reach and influence new audiences, and to broaden their own horizons.

Of course, travelling in first class there were still a number of privileged women, those who were wealthy and managed to remain largely unaffected by the economic downturn. They were the leaders of fashion, the international socialites at ease in the company of millionaires and royalty, whose names appeared in gossip columns and court circulars, and whose photos graced the newspapers and celebrity magazines so popular in this era. For them an ocean voyage in conditions of great luxury and opulence remained a necessary and welcome pause in their hectic social round. For the sophisticated woman of society, a transatlantic trip was still an opportunity for fine living, bringing perhaps a discreet romantic dalliance, conducted in what was effectively a giant floating hotel.

By relating the diverse experiences of a number of women, both seafarers and passengers of all classes, it is possible to illuminate the effect of the Depression on transatlantic travel in general, and to illustrate how individual women's lives were radically transformed by a mixture of necessity, aspiration and optimism.

In the early 1930s the number of women who worked on board transatlantic liners in any capacity consistently remained below 4 per cent of the total ship's company. These enterprising 'new women' of the modern age had defied social conventions of the era by electing to live independently. Though they might have a husband and children at home, as did Ann Runcie, or provide for a parent and siblings, like Violet Jessop, their motivation for seeking employment at sea was not solely a matter of financial pragmatism. There were compensations for the long hours, the discomfort, the 'Cunard feet', and those were to be found in the sense of adventure, of personal enterprise and the strong camaraderie often to be found below decks. They crossed the Atlantic every ten days primarily to make a living, but many of them regarded the grand ocean liner not merely as a workplace. It was a vast and complex entity, sophisticated, modern and progressive, and populated by a hierarchy of hundreds of skilled professionals. The ship's company were all united in a common purpose, to convey passengers around the globe, in safety and comfort, transporting them from one metropolis to another.

Before the Depression the transatlantic liners had been plying their trade across the ocean with nearly full capacity in both directions. It was a boom time for the travel industry, and women passengers of all classes benefited from the more comfortable conditions on board and the comparatively cheap fares for long-distance ocean voyages, while their seafaring sisters enjoyed ample employment in physically demanding but remunerative jobs afloat. But as the crisis deepened, wealthier passengers cancelled their plans for leisure and pleasure trips, and business travellers curbed their overseas visits, communicating with their commercial contacts overseas by letter or telegram rather than in person. Cut-throat competition drove ticket prices down by 20 per cent as companies struggled to fill their ships. Cunard cut its third-class transatlantic return ticket price from £20 to £16 per person, which meant that a 6,400-mile round trip from Britain to America now cost approximately one and a half pennies per mile.

With a surplus of tonnage, shipping companies were determined to create new markets. Some firms offered hastily organised pleasure cruises at unprecedentedly low prices, so that their ships could still generate some revenue, even if the profits were minimal. Leisure cruises offered economical trips on ocean-going vessels as holidays in themselves. 'Getting there is half the fun', Cunard's colourful posters trumpeted. 'Lust and sight-seeing' was the frank verdict of one observer. Whatever the personal motivation, sea travel was now promoted as an affordable, enjoyable experience, and the voyage as a welcome escape from the humdrum, workaday world.

There were important ramifications for female seafarers as the Depression intensified. Some women were able to cling to their jobs, grateful for a basic salary, even if the tips had dried up. There was now a different type of female passenger afloat, one who was more cautious and penny-pinching, often driven by economic necessity. Violet Jessop, a stewardess on the Red Star Line in the early 1930s, recalled that some of her more frugal female charges embarked on a succession of world cruises because it was cheaper than remaining at home. One character, whom she described as 'a fragile little old lady, a charming piece of Dresden', with carefully darned gloves – was waved off by her friends on a five-month round-the-world winter cruise. It was a lengthy, bargain holiday to avoid the worst of a British winter, having realised that by occupying a cheap, stuffy, inside cabin, she could live more comfortably and to a higher standard than would be possible ashore.

Violet counted herself lucky to retain her job as a stewardess. Many shipping companies were forced to economise on their staff costs, so they laid off their employees at short notice. There were 20,000 unemployed seafarers of both sexes in Britain in 1930, and by 1932 that figure had doubled. In each major British port there was usually a place where male seafarers would congregate between ships, in the hope of being recruited by the shipping companies for their next voyage. In Liverpool that place was a public house, the Dock Tavern, but female seafarers could not hang around in drinking dens looking for employment without causing a scandal and so they were obliged to apply in writing to the victualling department or lady superintendent of individual shipping lines, supplying references and evidence of their previous experience, in the hope of a hotly contested vacancy.

Those seafaring women whose jobs were dependent on the ships that conveyed migrants to the New World were the hardest hit. The economic downturn had caused unprecedentedly high levels of unemployment in the United States, and American immigration quotas were further squeezed. Edith Sowerbutts, veteran conductress of the North Atlantic run, found her maritime career brought to a juddering halt in the summer of 1931. Canada had finally halted assisted immigration, and so conductresses were no longer needed. There was no severance pay, only the wages she had earned on her last voyage, and of course she did not normally receive gratuities as her third-class clients were impoverished. Edith considered herself lucky to get a free trip back to Southampton from Antwerp, where she had been abruptly given her notice. Writing some fifty years later, she called her treatment 'shabby', and observed that conductresses and stewardesses always had inferior accommodation, few salary increases and little chance of promotion compared with their male counterparts. Nevertheless, she returned to terra firma with genuine regret:

> So, this was the end, then: no more unaccompanied women and children; no more happy, carefree days in New York

and Philadelphia ... It was farewell to delousing and bath sessions, immigration queries from cranky officials who had been forced to rise too early in the morning to meet incoming ships – it was adieu to sea breezes and force nine gales, and sayonara to starlight and gardenias ... And how one missed people, lots of people, all ages, types and nationalities. In my handbag was just the usual round-voyage payoff for thirty days. Atlantic conductresses were thus summarily cast off as their ships came into the home port, just like that! Each one of us, according to our age and qualifications, had to seek work ashore, and that in a country where unemployment was rampant – over 2 million out of work in Britain.[1]

Once again, Edith blew the dust off her loathed typewriter, and she spent the next three years working in a number of temporary or freelance jobs in London. Some of them were casual evening work, undertaken after already completing a tedious day in an office, but all of them were poorly paid. At least there was some administrative work available, and she could scrape a living, sufficient to keep the Sowerbutts family housed and fed. Edith chafed at being 'stranded on the beach', as seamen called the lay-off between voyages and hankered after the chance to return to her old, familiar shipboard life. However, for a number of her female compatriots the dire financial situation was the impetus that propelled them across the Atlantic for the first time.

Those who had grown up in the most deprived parts of the British Isles were spurred on by the complete paucity of employment at home and resolved to try their luck overseas. For them a one-way transatlantic journey was a gamble, a brave leap into the unknown. One such woman was Mary

Anne MacLeod, who emigrated from Scotland to the USA in 1930, celebrating her eighteenth birthday at sea on a liner heading for a new life in New York. Born in 1912 in the Outer Hebrides, on the Isle of Lewis, home for Mary Anne was the village of Tong, a huddle of cottages three miles north-east of Stornoway. She was the youngest of ten children; her father Malcolm was a fisherman and crofter, who also worked as a truancy officer for the local school, and her mother was already forty-four when Mary Anne was born. Life was harsh in this small rural community where the fishing industry was in decline. Raised in a Gaelic-speaking household, Mary Anne left school aged fourteen, presumably to help on the croft but she was keen to get away, and with three older sisters having already emigrated to the USA, either working or married, she had ample incentive to escape to the New World.

Certainly the Old World did not offer much; the population of Lewis were very poor, surviving through subsistence farming and herring fishing. The climate was harsh, and most people lived frugally in an inhospitable setting. The economic depression was exacerbated by the loss of many of Lewis's menfolk. More than 6,200 men from the Western Isles had gone to serve in the Great War, and 800 had been killed in action, more than one in eight. Of those who had survived the carnage of the trenches, a cruel tragedy struck in the early hours on New Year's Day 1919, within sight of home. Two hundred and eighty-four demobbed servicemen were packed on to a requisitioned yacht, the *Iolaire*, which was to carry them back to Stornoway, the capital of Lewis, after long years of war. The vessel set out on the last day of 1918, overladen with exuberant, homesick men, but it sailed into a howling winter storm, the crew lost their bearings and the yacht was wrecked on submerged rocks near the entrance to Stornoway Harbour. Two hundred and five desperate men drowned within metres of their homes, their bodies later washed up in the harbour and on the beaches where they had played as children. The sinking of the *Iolaire* was a further devastating blow to the people of Lewis, wiping out almost an entire generation of young men. As the *Scotsman* reported on 6 January 1919: 'The villages of Lewis are like places of the dead ... the homes of the island are full of lamentation – grief that cannot be comforted.' It was the worst peacetime shipping disaster in British coastal waters of the twentieth century.

Mary Anne was six when the *Iolaire* sank, and in subsequent years she was aware of the general shortage of eligible menfolk, and the surplus of 'spare women'. By the late 1920s the recession was biting hard on Lewis. Many young people left the islands to find work, often going to join relatives already in the United States or Canada, where they were assured jobs were waiting for them. Mary Anne applied for and received her immigration visa for America, number 26698, on 17 February 1930. She boarded the SS *Transylvania* from Glasgow on 2 May 1930, heading for a new life in the New World. Her vessel was a three-funnelled twin-propeller liner, an Anchor Line ship, which arrived in New York on 11 May, the day after Mary Anne's birthday; the passenger manifest shows a handwritten alteration to her given age, from seventeen to eighteen. A photo survives of her standing on deck on board the ship, wearing a sweater bearing the initials MM. She was five feet eight inches tall, with a fair complexion, fair hair and blue eyes, and had $50 in her purse.

On the *Transylvania*'s passenger list for all aliens (that is, anyone not already a US citizen) Mary Anne MacLeod made three declarations: first that she was moving to the USA permanently; second that it was her intention to seek American citizenship; and in the section asking 'whether alien intends to return to country whence he [*sic*] came' the answer she gave was no. She also declared that she would be living with her sister Mrs Catherine Reid, 3520 6th Avenue, Astoria, Long Island. In addition, she listed her occupation as a 'domestic', a catch-all description meaning some kind of household servant or maid, like her sisters. Despite the general tightening of American immigration rules for all nationalities throughout the 1920s, British-born servants were still very much in demand in the north-eastern states and cities in the 1930s. Indeed, the American immigration authorities had relaxed the annual quota on 1 July 1929, so that the number of emigrants from Great Britain and Northern Ireland entitled to enter the United States was increased briefly from 34,007 to 65,731, and Mary Anne was one of the beneficiaries.

In May 1930 the *Stornoway Gazette* reported:

> There is quite an exodus of young people, male and female, from this parish for Canada and the United States. Our straths and glens will soon be peopled only with middle-aged and elderly people. Most of these young people take kindly to the life of those distant lands but they never forget the 'old folk at home.' They leave home with a determination to succeed and because of their courage, endurance and reliability they are generally successful ... Several have left from here this week and we wish them *bon voyage*.

Mary Anne's decision to go to New York was partly as the result of a scandal that had struck the MacLeod family in 1920. Her older sister, Catherine Ann, born in 1897, also known as Kate or Katie, become pregnant out of wedlock, and was sent to Lanarkshire to have the baby. Her daughter, named Annie, was born in Airdrie on 5 December 1920. No father's name appeared on Annie's birth certificate, and when the new mother and child returned to Tong in 1921 Annie was handed over to be raised by the rest of the MacLeod family. Mary Anne, the youngest of the ten MacLeod siblings, was eight years old when her niece arrived and, as was often the case in large, poor families of the time, probably had an active role in taking care of Annie, playing with the little girl as she got older. Meanwhile, Catherine assessed her options. It would have been difficult for her to continue to live in Tong, a small, tight-knit community, where everyone knew about the baby, and speculation about the unknown father would have been rife. It was a conservative and rigid society, adhering to the strict beliefs and practices of the Free Presbyterian Church, and by the standards of the day Catherine was 'damaged goods'; no local man with any thoughts of respectability would marry her now, especially as there was a shortage of suitable men. It was time to try her luck overseas.

In 1921 she set out for New York to find work as a domestic servant, and a new start in life. In Manhattan Kate married a butler from Scotland, George A. Reid, on 26 March 1926. Four years later they were joined by her immigrant sister Mary Anne, also following a well-trodden family path, for two other elder sisters had emigrated to the USA in the 1920s, and found themselves expat husbands – Christina Matheson and Mary Joan Pauley, who had been in domestic service when she married an English footman, Victor Pauley.

It was thought that Mary Anne was staying with her sister as part of a 'holiday' in New York, and that she had no intention of emigrating to the States until she met her future husband. However, as the offspring of an impoverished crofter, it is highly unlikely she would have taken a vacation to New York; the more prosaic truth is that she had deliberately gone to America to seek work as a domestic servant, as her sisters had done successfully before her, and she declared as much on her immigration documents.

In 1930 Mary Anne met Frederick Christ Trump at a party. He was a property developer of German heritage, and ran a family firm with his formidable mother, Elizabeth. He was seven years older than Mary Anne, ambitious and on the make. The attraction was instant, and Fred told his mother the same night that he had met the woman he intended to marry. Fred Trump had been just twelve years old when his father died in the great Spanish flu epidemic of 1918, and he took over the family's real estate business, building houses. The stock market crash of 1929 occurred when Fred was twenty-four, and as a means of helping the American economy, President Roosevelt announced major public building programmes. Between 1935 and 1942 the Trump family's property empire boomed through those liberal economic initiatives.

In the summer of 1934 Mary Anne returned home to Scotland for three months, to see her family and perhaps to announce her engagement. Details are sketchy, but there was a Mary MacLeod from Tong, aged twenty-two, named on the incoming passenger list of the SS *Cameronia*, a ship that arrived in Glasgow on 11 June 1934. She is described as a nurse, and in fact there are a number of Mary MacLeods (with various spellings) from Lewis listed as crossing the Atlantic that summer, so the truth is elusive. (Though interestingly, her eldest daughter, Maryanne, mentioned many years later that their mother worked as a nanny during her early years in the States; it may be that she gave a different answer about her occupation in 1934 to match her most recent employment.) Nevertheless, it was definitely Mary Anne MacLeod who travelled back to New York on the SS *Cameronia*, arriving in the USA on 12 September 1934. She was already an American citizen in all but name, because she was travelling on a 're-entry permit' obtained from Washington on 3 March 1934. These prized and valuable permits were only granted to immigrants intending to stay in the States and become US citizens. Once again, her occupation was given as 'domestic', and the documentation stated she would be living with her sister Catherine Reid at Glen Head, Long Island. However, by April 1935, according to a later census of 1940, Mary Anne was living at 175/24 Devonshire Road in New York, the Trump family residence. This was a prosperous middle-class area of Long Island known as Jamaica, in the borough of Queens.

The wedding had taken place in January 1935; Frederick Christ Trump and Mary Anne MacLeod were married at the Madison Avenue Presbyterian Church, and held a stylish wedding reception for twenty-five guests at the Carlyle Hotel in Manhattan. The *Stornoway Gazette*, keen not to miss a detail ('Tong Girl Weds Abroad'), reported that the bride wore a 'princess gown of white satin with a long train and a tulle cap and veil', and her bouquet was of 'white orchids and lilies of the valley'. The couple honeymooned with a weekend in Atlantic City, but Fred's dedication to business was such that he was back at work on Monday, and Mary Anne commenced married life in Jamaica, Queens. Fred's real estate development business prospered and by 1940 the burgeoning Trump family were living in a house built by Fred, a large, red-brick, white-columned home positioned on top of a grassy hill. Mary Anne ran the house with the help of a live-in Irish-born maid, Janie Cassidy, also a naturalised citizen. Astute, attractive and charming, Mary Anne had a great deal of stamina and as the matriarch of a real estate business she was ambitious that the family should do well. Fred's business as a property developer was booming; he would buy land, build residential accommodation on it, then sell or rent out apartments to tenants.

Their first child, Maryanne, was born in April 1937, then Fred Junior in 1939; Elizabeth in 1942; Donald John in June 1946, and Robert in 1948. Mary Anne was very ill following the birth of Robert and her life hung in the balance for some time. On her eventual recovery, Mary Anne embarked upon a career in charity work, mostly around the family home on Long Island. She worked tirelessly, volunteering at a local hospital, and was actively involved with schools, charities and social clubs, while Fred worked closely with young Donald in the family's flourishing property empire. Mary Anne adopted an immaculate bouffant hairstyle and dressed smartly, and was rumoured to have her chauffeur drive her in her Rolls-Royce to pick up the bags of coins from the automated laundries the Trumps installed in the basements of their apartment blocks. Her son, Donald, attributed his own sense of showmanship to his mother, remarking in his book, *The Art of the Deal*, 'She always had a flair for the dramatic and the grand.' She would often return to her Scottish roots by visiting Tong, would attend the local church and easily reverted to speaking Gaelic to her relatives and the neighbours.

Fred Trump died in 1999, and Mary Anne in 2000. The *Stornoway Gazette* recorded her passing: 'Peacefully in New York on 7th August, Mary Anne Trump, aged 88 years. Daughter of the late Malcolm and Mary Macleod, 5 Tong. Much missed.' By crossing the Atlantic, Mary Anne left behind abject poverty in Scotland, and within twenty years she was a New World matriarch, the socially mobile wife of a wealthy real estate developer. She was a mother to five children, one of whom, Donald Trump, was to become President of the United States. Four of the MacLeod sisters were economic migrants; faced with financial hardship, social censure and lack of opportunity at home, they were prepared to take risks and embark on new lives on a distant continent, a long way from a tiny Scottish croft.

The possibility of permanent emigration to America or Canada remained an appealing prospect to many British-born women. Edith Sowerbutts, a globetrotter before she went to sea as a career, was very tempted to leave her unconvivial, poorly paid temping jobs in London and apply to emigrate to the USA, but decided against it because of her family commitments, particularly caring for her ailing and elderly mother. Those who felt themselves to be stuck in the economic mire of the early 1930s were particularly susceptible to the alluring image of life in America, as portrayed through sophisticated and engaging Hollywood movies, avidly watched in cinemas all over Britain. Accomplished screen performers such as Fred and Adele Astaire had enormous international appeal, and frequently took the Atlantic Ferry to fulfil professional engagements, performing in person in London or New York. Popular entertainers of all kinds were much in demand on both sides of the Atlantic, despite the economic strictures of the Depression. This was an era that saw many cross-cultural exchanges between Europe and the North American continent, from song-and-dance acts to the more highbrow pursuits. There was a great vogue for inviting female authors on lecture tours to promote their books, and to debate the pressing cultural issues of the day. British writers, such as Vera Brittain, were very popular in the United States between the wars, and they were aware that the opportunity to travel and reach new audiences could have a profound impact on their subsequent careers, not to mention their sales.

E.M. Delafield was the pen-name of Edmée Elizabeth Monica Dashwood, a writer who found fame and an appreciative audience with her humorous book, *The Diary of a Provincial Lady*. The book takes the form of a journal apparently kept by a rather insecure upper middle-class woman who is living with her family in rural Devon, trying to maintain standards despite being perennially short of money. Originally published in instalments in *Time and Tide* magazine, it appeared in book form in December 1930 and rapidly became a bestseller. In the sequel, *The Provincial Lady Goes Further*, the heroine becomes established as a writer and fulfils her dreams of making brief forays to London, to give 'talks', much to the resentment of Cook, the French governess, her two children and her taciturn husband, Robert. With the second instalment also proving popular, in 1933 Elizabeth Dashwood's American publisher, Cass Canfield of the firm Harper's, invited her to undertake a lecture tour of the USA and Canada. It was an exhausting itinerary: between 4 October and 2 December 1933 she travelled from New York to Chicago, Cleveland, Ohio, with a swift diversion to Toronto and Niagara in Canada, then on to Buffalo, Boston and Washington D.C. She encountered strangers and well-wishers, experienced wonderful generosity and occasional mutual cultural bafflement, visited Harlem and the mansions of the wealthy, attended endless parties and attempted to answer probing questions on the issues of the day. She was struck by the hospitality and warmth of her reception. Elizabeth Dashwood wanted to visit the family home of Louisa M. Alcott, author of *Little Women*, in Concord, and having been told that this would not be possible, called on the support of Alexander Woollcott, the renowned literary critic from the *New Yorker* magazine, and radio star, whom she had met over cocktails and sandwiches in Manhattan. Mr Woollcott replied that he would gladly 'mention it in a radio talk', if her visit to the Alcott House and Museum went ahead. The tour organisers were so impressed by her august media connections that they arranged for Elizabeth to have an exclusive tour of the house, which she found to be a fascinating insight into the world of her literary heroine.

The book that resulted from this tour, *The Provincial Lady in America*, a breathless journal, was published in 1934, and was based on the writer's own experiences, though of course it is a work of fiction. The central character's self-deprecating account of the pitfalls of life on board the ocean liner ship is amusing and acutely observed. Outward-bound, the Provincial Lady travels grandly in first class, is horribly seasick and relies on the kindness of her stewardess. Touted as a celebrity by her publishers, she is besieged by photographers and interviewers on arrival in New York. On her return journey, the literary tour completed, she finds that her publishers have booked her into third class. She is so glad to be almost home that when she finally catches sight of her taciturn husband Robert, who has come aboard the ship as it docks to meet her, she bursts into tears. He concedes that he has missed her, and pats her arm, a remarkable gesture for a stalwartly undemonstrative man.

Elizabeth Dashwood's lecture trip to America and Canada not only provided her with ample material for another popular book, it also revealed the mutual appreciation that largely existed in the early 1930s between the British and North American populace, particularly the more literary types living in the bigger cities. As a result of her literary tour and its successful novelisation, Elizabeth Dashwood was offered a very unusual project. Cass Canfield proposed that she should travel to Communist-controlled Russia, to spend six months working on a collective farm, in order to write a book on the subject. It was a dangerous and difficult time even to attempt to enter the Soviet Union: Josef Stalin was consolidating his power, a devastating man-made famine had been inflicted on the kulaks of the Ukraine, and political purges were rife. Nevertheless, Elizabeth was intrigued, especially as she already spoke some Russian. She sailed from London to Leningrad in the summer of 1936 and, despite obfuscation and suspicion from the Soviet authorities, 'Comrade Dashwood' found a billet with a farming commune where she underwent considerable privations.

When it was time to leave, travelling by ship from Odessa, she had to smuggle her 30,000-word manuscript out of the Soviet Union to prevent its confiscation. She ripped the cardboard covers off her notebook, and, with difficulty, managed to squeeze the pages down the back of her suspender belt, the elastic clamping the pages to her spine. The manuscript remained undetected by the Russian customs officer who was checking her cabin and her belongings for prohibited items.

On publication in 1937, *Straw Without Bricks: I Visit Soviet Russia*, as it was titled in Britain (in America it was *I Visit the Soviets: The Provincial Lady in Russia*), startled western reviewers; an author previously known for depicting the mild-mannered minutiae of bourgeois life had produced an authoritative first-person report on conditions in the secretive Soviet Union. But her account was recognised as authentic and truthful by Malcolm Muggeridge, the former Moscow correspondent of the *Manchester Guardian*, and E.M.

Delafield's professional reputation was enhanced; she had started out as a mildly nervous Provincial Lady, had become a cosmopolitan literary lecturer by travelling to and around America, and as a result became an eyewitness to a fascinating, secretive era in Russian history.

While E.M. Delafield recorded and commented on profound world events as she saw them, in 1934 another woman's transatlantic experience was to change history. A casual shipboard romance set off a chain of events that led eventually to the abdication of Edward VIII. The Prince of Wales, eldest son of King George V and Queen Mary, was the most famous bachelor in the world. He was a serial philanderer, and his preferred quarry was married women, especially glamorous and *soignée* Americans. In 1927 the prince dropped his long-standing mistress Mrs Freda Dudley Ward in favour of Thelma Morgan, the American-born wife of Viscount Marmaduke Furness. She was a beauty and an American heiress, whose twin sister Gloria had married Reggie Vanderbilt. The Morgan sisters, whose father came from the famous banking family, were cosmopolitan and inveterate transatlantic travellers all their lives.

Thelma soon became aware that 'Fiery' Furness, her irascible husband, was having affairs. Her bruised ego was soothed when the Prince of Wales invited her to cocktails at St James's Palace, followed by dinner and a dance. By December 1927 the prince and Thelma's names were beginning to be linked in the society pages of the newspapers. Thelma spent her evenings with the prince in fashionable London nightclubs, and they passed long weekends together at Fort Belvedere, the prince's home in Windsor Great Park. They entertained friends, she taught him *petit point*, he played the bagpipes while wearing his kilt, and pottered in the garden. It was all strangely domesticated, though Thelma noted that 'the outward shyness of the Prince masked a whim of iron'.[2]

Thelma's husband was complaisant, and the Furness marriage finally ended in a discreet divorce in 1933, on the grounds of his adultery. Marmaduke swiftly married again, but Thelma's status was now somewhat nebulous in an era of more rigid social protocol. She was an American, so held to be outside the definitions of rank in the British class system. This was her second divorce, and Thelma was nominally Catholic, though not apparently to the point of adhering to the Church's strict views on the indissolubility of marriage. Nevertheless she claimed subsequently that she held out no hope that she could marry the prince and in time become queen, as under the Act of Succession of 1701 no senior member of the British royal family could marry a Catholic without renouncing their rights to the throne. Meanwhile, she was independently wealthy, and felt assured of the prince's affection for her.

However, the warning signs were there. Thelma's predecessor, Freda Dudley Ward, had created a vacancy for 'married royal mistress' when she divorced her husband. The prince possibly suspected that Thelma as a divorcee might become either problematically needy, or, even worse, could be casting around for a third wealthy husband. Nevertheless, they attended balls and charity events together, and rented a large house in Biarritz for the summer so that they could entertain friends. Within court circles they were considered an established couple and Thelma's position as the prince's *maîtresse-en-titre* was acknowledged by the upper strata of society, while her discretion and tact were appreciated by the royal family. She appeared to present no threat to the established order, so it is ironic that it was Thelma herself who first brought Wallis Simpson into the orbit of the Prince of Wales.

In late 1930 or early 1931 her older sister Consuelo asked her if she could bring along an American acquaintance for cocktails at Thelma's house in Grosvenor Square that afternoon. Sociable Thelma agreed, and liked the new arrival, Mrs Wallis Simpson. Some more friends joined the little party, and the Prince of Wales dropped in. Thelma introduced the prince to Mrs Simpson, and in the following two years Wallis and her husband Ernest were gradually absorbed into the circle of friends around Thelma and the prince. Thelma came to regard Wallis as one of her best friends in England. The Simpsons first stayed at Fort Belvedere for the weekend in January 1932; as usual, Thelma played the hostess.

Thelma helped Wallis prepare to be presented at court, lending her a formal train and feathers. Meanwhile Thelma's domestic idyll with the prince continued; she organised the purchase of Christmas gifts for his staff, and wrapped each present with the help of the Simpsons. They put up a Christmas tree at Fort Belvedere, festooned by the prince with decorations. They attended Ascot together, and socialised with the prince's brothers, George and Bertie, and his young wife Elizabeth. They danced together to imported American records, and on one occasion Thelma and the Duchess of York skated together across a frozen pond, howling with laughter.

It was early January 1934 when Gloria invited her twin Thelma to join her in New York so that they could travel together to California for a holiday. Thelma was tempted by the thought of a sea voyage and a winter break, and told the prince she would be away for just five or six weeks. He seemed unhappy at the prospect of her absence, but she pressed on with her plans. A few days before her departure, Thelma invited her friend Wallis to lunch with her at the Ritz. Wallis said, 'Oh Thelma, the little man is going to be so lonely.' Thelma asked Wallis to look after the prince in her absence: her exact words apparently were, 'See that he does not get into any mischief.'

Thelma's farewell dinner with the prince at the Fort almost made her change her mind when the car appeared to take her straight to Southampton, as he looked so forlorn. 'I'll be back soon, darling,' she promised. She sailed on 20 January 1934 and returned to London on 22 March. What happened in the intervening two months was to have massive repercussions for Thelma, for the prince, and for the British royal family.

Thelma and the prince kept in touch through cables and lengthy international telephone calls; he even managed to reach her when she was visiting a film studio in Hollywood. After a sociable sojourn in California, mingling with old friends such as Louella Parsons, Marion Davies and William Randolph Hearst, Thelma returned to New York with time to spare before boarding the *Normandie* back to London and the prince. At a dinner party she was seated next to Aly Khan, the son of the fabulously wealthy Aga Khan, the leader of the Nizari Ismaili Muslims. Aly was a twenty-two-year-old bachelor, an international playboy and racehorse owner, and immensely charming. Aly Khan had some business matters to attend to, so he asked Thelma to cancel her booking and stay on in New York an extra ten days, so that they could travel together to England. She refused, but, undeterred, he invited her to dinner and they went dancing afterwards.

Arriving in her luxurious stateroom prior to sailing, Thelma was gratified to find it crammed with red roses, and affectionate notes from Aly. The ship set sail, and the following morning Thelma was having breakfast in bed when the phone rang; Aly was on the line, inviting her to lunch. To her shock, he announced he was also on board the ship; he had hurriedly completed his business affairs on the Eastern seaboad, had flown back to New York and joined the same ship. Over the following six days he laid siege to her and she was flattered; he was debonair, witty, generous and amorous. She dined with him every night of the voyage and she quickly realised she was out of her depth. She later described the interlude as a mere 'flirtation', though some accounts say that he proposed marriage, telling her that she could have no future with the prince. Thelma was a still-young divorcee, and unable to marry her Prince Charming for constitutional reasons, so Aly Khan might have offered her an alternative future. He was certainly keen; desperate not to let her go at the end of the voyage, Aly persuaded her to offer him a lift in her chauffeur-driven car from Southampton to London.

As they waited to disembark, Thelma received a long-distance call from the Prince of Wales, asking about her travel plans on landing – would she go straight to the Fort to have dinner with him? Thelma said she couldn't, that she had promised a lift to London to a friend. They arranged instead that the prince would come to her house for supper that evening, as soon as she returned to London.

Back in her London home, Thelma was reunited with the prince. But their conversation was stilted and awkward, not what she expected after eight weeks of absence. The prince suddenly volunteered, 'I hear Aly Khan has been very attentive to you.' Thelma was startled that he knew about her 'flirtation' on board the ship, and asked if he was jealous, but the prince would not answer. The evening petered out in further small talk, and Thelma was worried at the *froideur* of the man she had assumed adored her. He invited her to stay at the Fort for the weekend, but when she got there, she found him formally cordial, and anxious to avoid any intimacy.

It took Thelma twelve days to learn the truth, and during that time coy references started to appear in the gossip columns hinting at her shipboard dalliance. One gossip columnist reported that she was looking 'ten years younger' when she stepped ashore. Seeking reassurance, Thelma phoned her close friend Wallis Simpson, and went to her apartment in Bryanston Court to pour out her worries. What had changed in the two months she had been away? Wallis reassured her in anodyne tones that the prince had probably just missed her company. But then they were interrupted by the housemaid, Kane, calling Wallis to the phone in the next room – the Prince of Wales was on the line. Thelma heard Wallis say, 'Thelma is here,' and she expected him to ask to speak to her too, but no summons came. When Wallis returned, she made no reference to the call, and Thelma left, even more confused and alarmed.

The denouement came the following weekend during a house-party at Fort Belvedere. It was over dinner on the Saturday that a seemingly minor incident revealed the truth. The prince picked up a piece of lettuce with his fingers, and Wallis playfully slapped his hand. Thelma caught her eye and shook her head as if to warn against such familiarity; but Wallis held her gaze defiantly. Thelma realised the truth: 'Wallis – of all people,' she wrote.[3] Later, when Thelma was briefly alone with the prince, she asked him, 'Darling, is it Wallis?' His features froze and he told her not to be silly, leaving the room immediately. Thelma left Fort Belvedere the following morning, never to return.

Humiliated, Thelma found solace with Aly Khan in Paris. They travelled to Spain, accompanied by his valet and her maid. They motored to Barcelona, with Aly at the wheel, touching speeds of up to 100 mph. Thelma had burned her bridges now, and Aly was exciting, impetuous company. They moved on to Seville, returned briefly to London, then on to Paris, Ireland and Deauville, following the horse-racing circuit. Thelma stiffly denied a press rumour that she and Aly Khan were planning to marry, but she rented his villa at Deauville for the summer, complete with his retinue of Persian servants. They spent mornings on the beach, and evenings entertaining friends, or gambling at the Casino. It was a hedonistic, febrile summer, but within months Thelma realised that she didn't love Aly.

The crisis came in September 1934, and once again it was a transatlantic trip that forced the issue. Thelma was at a ball organised by Aly for his father the Aga Khan at Claridge's Hotel in London, when she was called to the phone in the early hours. It was her twin sister Gloria Vanderbilt, who had been separated from her eleven-year-old daughter by a pincer movement from her own mother and her sister-in-law, Mrs Whitney. A bitter custody battle for the child was about to erupt in the New York Supreme Court, and she needed her sister Thelma.

It was 3 a.m. in London and the *Empress of Britain* was sailing at 8 a.m. from Southampton. Determined to catch that ship, Thelma telephoned her slumbering household, and her servants hurriedly packed so that her long-suffering maid Elise could catch the early-morning boat train with madam's luggage. Meanwhile Thelma was driven through the night to Southampton and boarded the ship at dawn, still wearing her silver lamé evening gown.

Thelma arrived in New York a few days before the sensational case was heard by the New York Supreme Court. Various witnesses testified that Gloria Vanderbilt had affairs with men, but the evidence was inconclusive. However, Gloria's French maid, Maria Caillot, dropped a bombshell at the end of her testimony, claiming that she had seen Lady Milford-Haven kissing Mrs Vanderbilt in a bedroom in the Hotel Miramar in Cannes in 1929. The court erupted in shock and the judge cleared the room of press and public, 'in the interests of public decency'. The reporters raced to the public phones to transmit this sensational snippet to their editors. Gloria assured her brief, Mr Burkan, that there was absolutely no truth in Maria's story, and that afternoon the hearing continued, but without the press and spectators. Consequently, the innuendoes about Gloria's supposedly adventurous love life were only refuted in the court, not in public, and the many inconsistencies in Maria Caillot's testimony that emerged under cross-questioning went unreported by the media. The compromise verdict of the court was that Gloria Vanderbilt was to have custody of her own daughter only at weekends and for the month of July. For the rest of the time the little girl was to live with Mrs Whitney, her aunt.

Dejected, Thelma returned to her house in London alone. It was apparent that her relationship with Aly Khan had petered out while she had been away. Ironically, the original rift between Thelma and her prince had occurred because of her 'flirtation' with Aly while she was on a transatlantic voyage, but her subsequent dash to the States months later in support of her family had given Aly the chance to become romantically occupied elsewhere. Their relationship limped on a little longer, but he was named on 18 June 1935 as co-respondent in a divorce case between Tory MP Loel

Guinness and his wife Joan. In a further public humiliation, on 6 August 1935, the Prince of Wales arrived in Cannes on a hired yacht for a holiday, bringing with him a crowd of friends, a terrier and Wallis Simpson, though not Ernest, who had tactfully declined the invitation. The party spent an idyllic two months sailing round the Mediterranean, and it was known in the innermost circles that the prince was now infatuated with Wallis, his previous lover's erstwhile best friend.

In January 1936 George V died and his eldest son, the Prince of Wales, became Edward VIII. Within eleven months the new king, Thelma's former lover, renounced his throne in order to be with 'the woman I love', Wallis Simpson. Thelma wrote: 'He had been the Prince Charming of the Empire, a man everybody loved … it seems to me he should have known that the British Empire could not and would not accept as their King a man who deliberately flouted the most deeply rooted traditions of Church and State.'[4]

Thelma was aware that her long-standing relationship with the prince was deliberately sabotaged by someone unknown to her. Perhaps she had been tailed by a detective, or reported by a well-connected fellow traveller, or maybe even an officer or a crew member aboard the ship. Whatever the source, the heir to the throne obviously knew about her brief shipboard romance with a practised playboy within hours of her return. Just like Thelma, Wallis Simpson was a twice-married American woman with an eventful past when she first met the prince, and that was part of her appeal. However, unlike Wallis, Thelma had been caught out in a 'flirtation' with another man, Aly Khan, seduced by the romance of the setting and the ardour of his pursuit while on a transatlantic voyage, and this was unacceptable to the prince.

There is a certain level of hypocrisy evident in this cautionary tale. The Prince of Wales routinely targeted married women whenever he had the chance, cuckolding their husbands, whom he regarded as friends. Members of his social circle accepted the double standards of the day: men were expected to philander, but women were required to remain faithful to their lovers, if not their husbands. If her shipboard dalliance with Aly had never happened, or had remained a secret, Thelma and the prince might have continued their discreet relationship, at least until he assumed the throne on his father's death less than eighteen months later, and possibly longer. If the prince had not sought solace with Wallis Simpson, following Thelma's ocean romance with another man, there would have been no grounds for his abdication. The British royal family might now be completely different, if Thelma had not succumbed to Aly Khan's charms.

There is rather a poignant postscript to Thelma's story. When she collapsed suddenly and died in the street in New York in 1970, in her handbag was found a battered miniature teddy bear, a keepsake from the prince who never became king. Thelma had taken the love-token with her everywhere, for nearly four decades.

Thelma Furness, wealthy heiress and chic socialite, royal mistress and frequent ocean voyager, was not ruined financially by the Wall Street crash and its aftermath. But she was twice-divorced, and concerned about her long-term emotional wellbeing and marital status. The two return transatlantic journeys she made in 1934 were to have dramatic consequences for her own future happiness. Hardworking female artistes such as Adele Astaire performed in theatres in the States and London, and for her and her dance partner and brother Fred, frequent travel on the Ocean Greyhounds – a popular term for the latest, fastest commercial lines travelling on the highly competitive North Atlantic run – was a necessary aspect of their professional life. Creative and successful writers, such as E.M. Delafield, recognised that the cultural Zeitgeist of the early 1930s offered them the chance to cross the Atlantic and explore America through the means of a literary tour, while the realities of the Depression spurred them on to boost their sales figures on both continents. Canny economic migrant Mary Anne MacLeod seized the chance to exchange an impoverished Scottish island for the heady possibilities of New York. For the pragmatic women whose livelihood relied on the great ships of the inter-war era, the Depression was an ordeal to be survived. Violet Jessop adapted to the new realities, and worked as a stewardess on cruises for a minimal salary, serving genteel passengers who were often too cash-strapped to augment her pay with tips. Edith Sowerbutts's services as a conductress, escorting would-be immigrants, had been dispensed with in 1931, and for a few years she was forced to find a less convivial occupation ashore. For each of these women, during the Depression years of the early 1930s, their lives and their futures were moulded by transatlantic travel on the great ships.

By 1934 the world economy was starting to recover; transatlantic trade had begun to pick up, and women were increasingly employed once again on board ships, often in more diverse and responsible positions. In particular, trained nurses were engaged for the biggest and newest passenger liners. The invaluable services provided by VADs and Queen Alexandra's nurses on hospital ships in the Great War had proved the worth of nurses in critical care, and the merchant navy recognised the value of employing them on passenger ships. Their role was to assist the ship's doctor, and as well as general nursing skills they needed proficiency in assisting at surgical operations and providing post-operative care. Some ships seemed prone to having medical emergencies: the *Georgic* was known by Edith Sowerbutts as 'the appendicitis ship' because she was aware of at least three cases that had occurred on it, two of them fatal. Any operation carried a risk, especially in an era before antibiotics.

As the ships became more sophisticated, operating theatres were installed on board for emergency treatment, and a number of hospital beds for inpatients. By 1931 the *Majestic*'s medical quarters had sixty-two beds housed in eight hospital wards, and a large operating theatre. It was headed by Dr Beaumont, who had crossed the Atlantic more than a thousand times, and he had five male attendants and two female nurses.

In addition to caring for the convalescents, treating minor ailments in the clinic or attending to ill passengers in their cabins, nurses would also accompany the ship's doctor on his daily rounds inspecting the third-class passengers. On the modern liners the boundary between first class and third was very apparent. After descending several levels, from first class through second class to the lower decks of the ship, the doctor and nurse would pass through a discreet connecting door on a corridor, which was normally kept firmly locked. On the second-class side, the door appeared to be of carved oak; on the third-class side, the same door was utilitarian metal, covered in white enamelled paint, and it led into a long corridor, a functional but cramped space where the third-class children often played.

In the inter-war years, many professional nurses preferred working at sea to staying on land, as they had more autonomy and lighter duties afloat, far from the overbearing surveillance of hospital matrons. Their shipboard careers could last well into their fifties and sixties, as the daily workload was much less demanding than the constant physical activity demanded of a nurse in a busy hospital ashore. In addition, they had respect, as well as better pay and the excitement of travel. Nurses enjoyed far higher social status than stewardesses within the ship's hierarchy. The 'sister' had many of the privileges of the male officers, with her own cabin and the attendance of a stewardess, who would deal with her laundry and bring her early morning tea. She was able to dine in the first-class salon and was allowed to participate in all the ship's many recreational activities. She was not allowed to accept tips, but she could be given presents, and was often popular as a dance partner at evening parties.

Nurses' image as 'ministering angels' tended to made them very popular with men, both the passengers and crew, and it is telling that one stock heroine of the romantic literature of the 1930s is the 'floating sister'. The highly successful novel *Luxury Liner*, by Gina Kaus, published in 1932, is set on a fictional transatlantic vessel, the *Columbia*, sailing from Bremerhaven to New York. The hero is a doctor, Thomas, whose wife has run off with her lover. Thomas has been engaged as the ship's surgeon, and is travelling to America determined to win back his errant wife, who is also aboard the *Columbia*. However, Thomas falls in love with the ship's nurse, the apparently saintly and beautiful Sister Martha.

Her life has been blighted by a terrible tragedy, which has caused her to dedicate herself to caring for others on big ships. Like much romantic fiction, the central characters in this book incorporated popular role models of the time – a tale about a doctor and a nurse falling in love in the glamorous setting of an ocean liner would capture the reader's imagination.

Romance aside, in reality medical emergencies often had to be dealt with while the ship was far from land, and sometimes passengers died on board, a fact that was tactfully kept from their fellow travellers. During the 1930s Violet Jessop worked as a stewardess on many White Star Line world cruises, which started in New York, crossed the Atlantic and toured the Mediterranean. She had one particularly objectionable regular client, an elderly, wealthy woman known as the Baroness because of her imperious manner. An eccentric, she always travelled with her collection of live canaries, which were kept in cages in her cabin. She wore fabulous jewellery, but her hands and nails were often filthy. She was obese and obsessed with food, and she frequently complained to the chief steward, who ordered the crew to give her anything to keep her happy. Although she was a fractious character, often suspicious of her fellow passengers, the Baroness could also be extremely generous. On an earlier voyage, she was so upset by the accidental death of one of the ship's sailors, who was washed overboard off the coast of California, that she started a fund for his bereaved family and contributed to it generously.

On this particular cruise, when the ship docked at Cairo the Baroness went ashore sightseeing, but she returned with a temperature of 103 degrees, and was nursed in the ship's hospital. Three days later she died, and Violet and the nurse had to embalm her body and lay it out. Empty coffins were carried in the hold by all liners as a matter of course, but the Baroness was too fat to fit, so the ship's carpenter was summoned and hastily began to construct an extra-large casket. Because of the delay while he tackled the task, the Baroness's body developed partial rigor mortis, with one arm rising as if in protest. Having manoeuvred the body into the casket, and covered it with a white sheet, the crew left it in the alleyway of the hospital to await the captain's formal inspection. The news of the Baroness's death was still secret, so when a well-intentioned stewardess popped into the hospital with a treat for one of the patients, she had no idea what was in the large white-draped rectangle in the corridor on which she had stubbed her toe. She pulled aside the sheet, and was horrified as the corpse's arm shot up like a fascist salute.

Nursery staff were a welcome and necessary addition to the ship's roster of female employees in the early to mid-1930s. Shipping companies needed to make their vessels more appealing to the female passenger by providing dedicated facilities for children and staff to look after them. Nursery stewardesses were in charge of the facilities in purpose-built playrooms; they wore distinctive grey dresses with white collars and cuffs and white caps. Qualified nursery nurses, who had a higher status, were also now employed to provide daytime respite to families travelling with young children. On the North Atlantic run, where a journey could easily take a week, and the family cabins were often compact to the point of claustrophobia, harried parents were glad to deposit their energetic offspring for a few hours in the playrooms, which operated as a combined crèche and kindergarten. The children would have the company of other minors, and could amuse themselves safely under the watchful eyes of the nursery nurses. On the White Star liner *Regina*, the playroom was adjacent to the lounge, the reading room and writing room, in the centre of the ship. This facility allowed mothers and fathers to use the public rooms to chat, read, play cards, write letters or have a cocktail, yet still be within call of their youngsters. Babies and toddlers were cared for, changed, entertained and fed at regular intervals by the nursery nurses. For older children there were toys on offer – jigsaws, doll's houses, play houses, swings and rocking horses. The playroom was a godsend for women travelling with children, though canny fellow passengers knew to book cabins as far away from it as possible.

Having consigned her precious offspring to the care of a professional nursery nurse, the female passenger with a few precious hours to herself might well seek out the expertise of other women working on board the ship. Shopping might appeal; the *Aquitania*'s row of fashionable bijou shops was known as the Atlantic Rue de la Paix, and there attentive female shop assistants would dance attendance on lady customers. Clothes were much on the mind of the well-off transatlantic passenger; there were skilled seamstresses and fine laundresses on board who could work wonders with one's wardrobe, making repairs and alterations, or tackling stubborn stains. Turkish bath attendants and masseuses were available for body maintenance; in an era where it was fashionable to be slim and to look athletic, an hour in a steam room followed by a therapeutic pummelling from a woman who had seen every possible body shape and was sworn to discretion was an appealing prospect. And then of course there was the essential 'lady hairdresser', the Mistress of the Marcel Wave. Female hairdressers were sometimes called 'barbereens' in this era, and they gradually expanded into offering their clients beauty treatments too, such as manicures, pedicures and the application of make-up. It was a rewarding and interesting career for an enterprising and energetic young woman.

Kathleen Glendinning from Wallasey was featured in an article in the *White Star Magazine* of March 1934. She had worked as a hairdresser for seven years, travelling between Liverpool and the American ports, and had covered more than half a million miles in liners. In February 1934 she sailed to Boston in order to marry Ronald Crawford, who was employed by the White Star office there. The couple had met five years before, on his outward journey, and the romance had presumably been conducted at long distance. Kathleen Glendinning was photographed for the article, shaking hands on deck with the *Adriatic*'s Captain Freeman. She is wearing a smart fur coat with a shawl collar and a chic little hat; she looks glossy, smart and competent, heading west across the Atlantic yet again, this time into a new life.

That same year also brought a substantial improvement to Edith Sowerbutts's career. Three years after being laid off from her job as a conductress, she was working in the typing pool at the *Daily Express* in London, hammering out marketing letters, but her luck was about to change. The maverick proprietor of the newspaper, Lord Beaverbrook, appeared in the women's office on a surprise tour of inspection, and promptly sacked them all. Edith recalled: 'I thought His Lordship a typical tycoon, frog-faced in appearance, peremptory in manner. But he did me a good turn that day.'[5] She left the office elated, clutching her salary to date as well as two weeks' wages in lieu of notice. This modest windfall was enough to allow her to change her career, because now she could afford to equip herself with a new uniform. Her previous experience as a conductress was a considerable advantage, and this time she was lucky, there was a vacancy. Within a week she was engaged to work on the *Olympic*, the sister ship of the *Titanic*, on the transatlantic run.

Lord Beaverbrook's whim enabled Edith to buy the new uniforms she needed: striped blue and white morning dresses, dark blue afternoon frocks and large white aprons. Stewardesses also wore belts, cuffs, collars and the hated Sister Dora caps. Edith and her sister Dorothy, a stewardess with the same firm, came under the control of the chief lady superintendent, Miss Somerville, a former Cunard conductress. Miss Somerville had considerable authority within the company, as she and her two able assistants, Miss Moseley and Miss Prescott, selected and managed all sea-going women for Cunard and White Star, following the merger in 1934. The superintendents operated from Southampton and Liverpool, and one of them always came aboard whenever a ship docked or sailed, to meet the female crew and address any concerns. Edith was now in her late thirties, and she felt it was only a matter of time before she was deemed to be too old for junior office jobs on land, and was left without the prospect of any casual work. However, for stewardesses, a few extra years conferred the welcome advantage of added gravitas, and her previous experience as a conductress was valued by her female bosses. Edith also appreciated the collaborative nature and managerial skills of Miss Somerville and her deputies, feeling she could confide in them with any problems. They were to be immensely supportive of Edith and her sister Dorothy when they had family problems in later years. Meanwhile, Edith was happy to find herself back on the Atlantic Ferry, in a new role. As a stewardess, rather than a conductress, she had far fewer weighty responsibilities, and the prospect of earning decent money at last.

The year 1934 proved to be a watershed for the future of British-owned passenger shipping, because it was when work resumed on a vital project, the construction of the *Queen Mary*. Drastic action had long been required. In the mighty shipyards of Britain, the Great Depression had led to the lay-off of thousands of shipbuilders. Cunard's latest flagship, known at this point only as Hull 534, had been under construction at John Brown's in Clydebank since 1930. But by Christmas 1931 the outlook for the company's North Atlantic trade revenues was so grave that work was stopped, 2,000 ship workers were laid off from Hull 534, and for the next two years the project languished.

The inert hulk of the ship, which was to have been longer, bigger and faster than any other in the world, towered over the boatyard and adjacent workers' houses like a gigantic, rusty rebuke. The collapse of traditional industries that had made Britain the workshop of the world – iron, steel, coal, textiles, shipbuilding – came as a profound shock to the national psyche. Public interest was intense, with letters pouring in to Cunard's Liverpool offices, some even enclosing donations. British morale and international prestige were so bound up in Hull 534 that questions were asked in Parliament. An urgent confidential enquiry into the trading and financial position of British shipping companies involved in the North Atlantic recommended a merger of the two main firms competing for transatlantic passenger trade. In 1933 a new company, Cunard White Star Limited, was formed. The merger led to a review of the combined fleet and the subsequent sale overseas of surplus vessels. The rationalisation was long overdue, and in return the Treasury effectively underwrote the completion of Hull 534, to ensure that the new shipping company remained in British ownership. The new ship eventually cost £3.5 million to construct, approximately £225 million today. It had proved to be too big to be allowed to fail; the government agreed to support the completion of the project in order to alleviate the crushing unemployment in the area, and work resumed in April 1934.

Hull 534 was inadvertently named in honour of the consort of King George V, Queen Mary, and broke Cunard's ninety-year tradition of giving each vessel a resonant name ending in 'ia', after ancient Roman provinces. The company had planned to call this ship Victoria after the nineteenth-century British monarch and Empress of India, while still maintaining their brand identity. When a personal delegation from Cunard's board tactfully requested the king's permission to name the liner after 'Britain's greatest queen', he broke in immediately to say that his wife, Queen Mary, would be delighted to accept the honour. No one present had the nerve to correct the peppery though dim monarch. Subsequently, Cunard stoutly claimed publicly that it had always been their intention to name the ship after the current queen. However, Felix Morley, editor of the *Washington Post*, who sailed on the ship's maiden voyage in 1936, was told about the misunderstanding by Sir Percy Bates, chairman of Cunard, on condition that the truth wasn't published during Sir Percy's lifetime.

The launching ceremony of RMS *Queen Mary* occurred on 26 September 1934, attended by the king and queen. The half-built vessel was the largest ever launched into the narrow River Clyde, and there were serious safety concerns since the engines had not yet been installed. As 35,000 tonnes of metal accelerated down the slipway, drag chains attached to the hull acted as brakes. So vast was the ship that it caused a huge displacement wave on entering the river, swamping the footwear of tens of thousands of cheering onlookers standing on the river banks. It was all very British; even the weather had lived up to expectations. 'Crowds' Enthusiasm Unquenched by Drenching Rain' was the headline in the next day's *Daily Telegraph*, which published a sixteen-page supplement celebrating the occasion, penned by the paper's shipping correspondent, Hector C. Bywater. Two more years of finishing work lay ahead for the *Queen Mary*, but the project provided a resurgence for British industry as well as a much needed boost for public morale. When the ship finally left the Clyde for its sea trials in March 1936, 1.5 million people turned out to see it depart, so keen were they to witness this symbol of national pride. For many, the *Queen Mary* had come to embody the triumph of dogged determination and willpower over the enervating effects of the Great Depression. 'This great ship, freighted with the hopes of a nation' as Hector Bywater put it.

8

The Slide to War

———

On 27 May 1936 the *Queen Mary* left Southampton for New York, via Cherbourg. This was the vessel's maiden voyage, and nearly one million observers watched it inch out into the Solent on its inaugural journey, accompanied by a flotilla of smaller ships and boats. So great was the excitement that 15,000 people had paid five shillings a head to visit the ship in the week before it departed on its maiden voyage. Newsreel footage of the era shows crowds of excited visitors touring the ship, particularly families, with children in their 'Sunday Best' outfits, their fathers sporting suits and overcoats, and their mothers wearing hats and fur stoles. A trip around the *Queen Mary* before it sailed was a thrill, even if the visitors were not actually passengers, and were 'gonged off' by stewards when it was time for them to return to shore. The money raised went to charity, but Cunard had not anticipated the public's subliminal feeling that they had a right to own a piece of this mighty vessel; by the eve of sailing, every ashtray on board had been stolen as a souvenir.

Many had anticipated the launch of the *Queen Mary* for years, and now it was sailing day at last; passengers trotted up the gangways, to be directed to their cabins by officers and stewards. Baggage handlers took charge of tons of intricately labelled luggage, while armloads of bouquets were delivered by bellboys to the waiting arms of stewardesses. Among the floral tributes, a five-foot-long replica of the vessel, constructed out of white flowers, stole the show as it was carefully carried aboard. As the time of departure neared, most of the 2,079 passengers on board milled about on deck, hoping to catch a sight of their nearest and dearest down below on the quayside, so that they could wave frantically at each other, while the 1,100 crew readied the ship for its inaugural voyage. The band struck up for the umpteenth time, the last hawsers were cast off, and the great ship almost imperceptibly inched away from dry land. Small tugs adeptly nosed the *Queen Mary* out of the basin and into the Solent. In more open water, there was an armada of craft of all types and sizes – excursion paddle-steamers hired for the day, pleasure craft, fishing boats, naval cutters, yachts and dinghies. All were full of cheering onlookers, keen to provide a personal escort. Overhead, planes buzzed the ship, with news cameramen on board, recording the spectacular event for newsreels to be shown in cinemas, where it was watched avidly and triumphantly by millions. 'To Britons, she represents the restoration of Britain's supremacy on the seas. With her goes the hope and pride of a nation,' boomed the commentator above the swelling chords of 'Rule Britannia'. Among the passengers was a fourteen-year-old British girl called Heather Beagley, travelling to New York with her family. She recalled the 'tremendous excitement' of the voyage, and likened it to 'going to the moon now'.[1]

All classes were free to mingle on deck on this first voyage, and afternoons afloat were often spent in reclining chairs in the open air, with a rug over the knees, while an attentive steward hovered with a trayful of tea and cakes. The sumptuous interiors of the ship conjured up a fantasy world, especially in the evenings. As an entity, the ship most resembled a vast, floating hotel, but there was an atmosphere of great gaiety, with a constant kaleidoscope of galas, ballgowns, champagne and fine dining. Music for every occasion was provided by Henry Hall's Dance Band. The team of on-board photographers worked in shifts every night to meet the huge demand for commemorative pictures from party-going passengers.

There were hopes that the *Queen Mary* might arrive in New York in time to claim the Blue Riband, the much contested award for the fastest cross-Atlantic trip, but the ship was delayed by fog and missed beating the record, held by the *Normandie*, by a mere forty minutes. Nevertheless, the arrival of the *Queen Mary* in New York Harbour on 1 June 1936 was rapturous, as a flotilla of vessels of all types welcomed Cunard's new flagship, and the quayside was packed with cheering well-wishers. Heather Beagley recalled that aeroplanes flew alongside and overhead in salute, and fire hoses played over the Hudson River, casting rainbows in the sunshine with their arcs of spray. The pandemonium raised by the competing sirens, plane engines, bands and well-wishers' whistles and whoops was drowned out by the *Queen Mary*'s sonorous horn, a resonant vibrato subsequently likened by Edith Sowerbutts to hearing Dame Clara Butt singing 'There'll Always Be an England'.

Cunard's flagship vessel was intended to dwarf all other contenders on the lucrative and prestigious transatlantic route. The *Queen Mary* combined the finest classical marine architecture with advanced engineering expertise. Cunard believed that approximately 70 per cent of this ship's revenue would come from the American market. Consequently the interiors were reassuringly traditional, although leavened with a certain amount of whimsy and a restrained form of modernism. The veteran architect Arthur Davis, who had designed the interiors of the *Aquitania*, the *Franconia* and *Laconia*, had once more been engaged in 1935 for this new ship; in fact, this was his last major commission. According to his daughter, Ann Davis Thomas, speaking in 2004, 'he finally did the *Queen Mary*, which he hated. He was made to do it in Art Deco style, and it was not his bag. But the curious thing is, somehow, he did it. He managed to do it, and it's become almost a prototype of Art Deco.'[2]

The *Queen Mary*'s interiors incorporated subtle, natural materials and fine finishes. More than fifty tons of wooden veneers, taken from fifty-six different species of trees, represented all the countries of the British Empire. The accommodation ranged from luxurious staterooms to the more compact but well-designed third-class cabins. Cunard claimed that the public rooms represented 'those fundamental characteristics of British homes which are generally admired and appreciated by men and women of all nations'. However, when Cecil Beaton travelled on the *Queen Mary* on its maiden voyage, he criticised the lack of theatricality in the ship's interior, particularly the absence of a spectacular staircase suitable for making the *grande descente*. 'When constructing a boat, even a luxury liner, the English do not consider their women very carefully. There are hardly any large mirrors in the general rooms, no great flight of stairs for the ladies to make an entrance.'[3]

Women were more actively involved in the creation of the *Queen Mary* than in any previous British-built ships. The role of female artists, interior designers and artisans working on the preparation of the *Queen Mary* was acknowledged in the press, especially in an article in the London *Evening Standard* on 11 February 1936. Sisters Doris and Anna Zinkeisen provided numerous paintings for the prestigious

Verandah Grill and the white and gold Ballroom, while Dame Laura Knight created a special picture for one of the private dining rooms. Lady Hilton Young's marble plaque of Queen Mary, set in a panel of special burr walnut, was to be installed at the head of the main staircase facing the Main Hall. Meanwhile, Hetty Perry provided a decorative map, to be hung under the clock in the tourist smoking room, as well as wall decorations for the children's attractive playrooms in the tourist section. Margot Gilbert also painted a sequence on the theme of 'Dancing Through the Ages', representing the art form from stately measures and classical dances down to jazz.

The same newspaper article also recorded the role of thousands of unknown women working in manufacturing, engaged in provisioning and fitting out the great ship: 'busy fingers are plying needles and machines in Glasgow and Liverpool, in London and Ireland. Women old and young are putting every endeavour into their skilled tasks. Women operators are engaged in preparing fabrics and making upholstery; women's handiwork is being employed in making pillow-cases and coverlets, and even in preparing compass equipment, the latter a most delicate task.'[4]

The *Queen Mary* was luxuriously fitted out to meet every possible need. There were two indoor swimming pools, beauty salons, libraries and children's nurseries for all three classes, a music studio and lecture hall, a telephone system that could connect passengers to anywhere in the world, outdoor tennis courts, a squash court and dog kennels. There was an arcade of shops, with twenty-four large window displays, including haberdashery, gifts and tailoring. A fountain, deep sofas and baskets of fresh flowers enhanced the air of relaxed luxury. The ship was also the first transatlantic liner to have a purpose-built Jewish prayer room on board, available to all classes of passenger – an enlightened response to the growing mood of anti-Semitism in parts of mainland Europe.

The largest single space on board was the first-class main dining room (the Grand Salon), which was three storeys in height. At one end was a large map of the transatlantic crossing, with twin tracks representing both the more northerly summer/autumn route, and the winter/spring route (further south to avoid icebergs). During each crossing, a motorised model of *Queen Mary* would indicate the vessel's daily progress. An even more exclusive alternative to the main dining room was available to cabin-class passengers – the Verandah Grill on the sun deck. This à la carte restaurant could seat approximately eighty diners, and it was converted to the Starlight Club every night after dinner. Another popular spot for elite socialising was the Observation Bar, an art deco-styled lounge with wide ocean views.

Cunard's marketing department promoted the vessel's modernity, its technological sophistication and, above all, its breathtakingly vast scale. *The Queen Mary – A Book of Comparisons* was published in 1936 listing newsworthy statistics about the longest ship in the world. Modernist cartoons in red and black, like a Futurist version of the *Beano*, illustrated a plethora of impressive facts and figures. The engines generated a mighty 200,000 horsepower, while the ship, at 1,018 feet in length, exceeded the height of the Eiffel Tower (984 feet), the Pyramid of Cheops in Egypt (461 feet) and Westminster Tower, the home of Big Ben (310 feet). There were three acres of deck space given over to recreation, and twenty-one lifts, while the ballroom was the largest ever created on a ship. No detail was too prosaic: over 10 million rivets were used to construct the ship, and 30,000 electric lamps illuminated it, while six miles of carpet were cleaned every day with scores of vacuum cleaners.

Feeding the huge numbers of passengers and crew required victualling on a vast scale. Some of the staple foods apparently carried on an Atlantic round trip included 1,000 pineapples, 50,000 lbs potatoes, 3,600 lbs cheese, 3,600 lbs butter, 6 tons fresh fish, 60,000 eggs, 20 tons meat, 12,800 lbs sugar, 3,600 quarts milk, 1,200 lbs coffee, 2,000 quarts ice cream, 200 boxes of apples, 280 barrels of flour and 5 tons ham and bacon. The kitchens covered an acre, and over 40,000 meals would be served during a single voyage, using 500,000 pieces of china, glassware and table silver. On-board linen supplies included 210,000 towels, 30,000 sheets, 31,000 pillow cases, 21,000 tablecloths and 92,000 napkins. In addition, the *Queen Mary* carried 5,000 bottles of spirits, 40,000 bottles of beer, 10,000 bottles of table wine, 60,000 bottles of mineral water, 6,000 gallons of draught ale, 5,000 cigars and 20,000 packets of cigarettes on each transatlantic voyage.

In the 1930s ocean liners captured the public imagination in the way that aircraft and spacecraft were to appeal to later generations. The perception of technological rivalry with the ships of other nations was bound up in this appeal; the French had launched the superlative *Normandie* the previous year, and inevitably the two super-liners were compared.

The *Normandie* had entered service in 1935. It was daringly avant-garde, and it re-established France as the pre-eminent nation for visual and material culture. Compagnie Générale Transatlantique (CGT, and typically known overseas as the French Line) had commissioned the best French designers and artists to provide chic settings of unimaginable luxury and modernity. The *Normandie* was the biggest ship afloat when it was launched, and the largest art deco object ever created, according to Ghislaine Wood, co-curator of the Victoria and Albert Museum exhibition, Ocean Liners. It was breath-taking in its audacious approach to interior design. The ship was full of huge public spaces, designed like the most daring theatrical or movie sets, with dramatic entrances, mirrored surfaces, framing doorways, ornate screens, settings where the passengers could pose and 'grandstand'. The first-class dining saloon was nearly 300 feet long, rising through three decks to accommodate 700 diners. The entrance boasted bronze doors twenty feet high, and the decorative scheme was a medley of onyx, gold and crystal, with illuminated glass fountains and lights by Lalique.

There were many opportunities for the woman of fashion to make a spectacular *grande descente*, as vast, decorative mirror-lined staircases linked the various public rooms of one floor to another, acting as catwalks for the well-dressed passengers.

The ship was heralded as a triumph of the modern age and instantly became the pride of the French Line. An estimated 100,000 spectators had lined New York Harbour for its triumphant arrival. Its passenger list included Ernest Hemingway, Marlene Dietrich, Walt Disney, Salvador Dali, Douglas Fairbanks Jr, James Stewart and Bing Crosby. The *Normandie* was chic, sleek and fast, securing the prestigious Blue Riband on its inaugural transatlantic crossing. Despite the *Normandie*'s understandable appeal to the international fashionable set, to celebrities and Hollywood moguls, maharajahs and millionaires, the ship was not commercially successful in its own right. CGT ran it mostly at a loss, subsidised by rather more modest vessels of the same line, despite the iconic advertising poster by Cassandre.

Just eleven years after the Armistice, the German merchant fleet were once again serious rivals in transatlantic trade. In July 1929 the German liner *Bremen* sailed on its maiden voyage to New York and won the coveted Blue Riband with an average of 27.38 knots. The *Bremen*'s sister ship, the *Europa*, also broke speed records in transatlantic crossings in 1930. The *Europa* and the *Bremen* were the twin transatlantic flagships running between German ports and North America. Noël Coward was familiar with both, having crossed the Atlantic on one of them (he claimed he could not tell them apart) in the spring of 1932. He later wrote about his vague sense of suppressed guilt in patronising a German-owned ship, even though the Great War, which had menaced his teenage years, seemed so far in the past. Despite the apparent sense of equanimity among the European nations in the early 1930s, he maintained a lingering sense of unease about the future.

Sir Percy Bates, Cunard chairman, noted as early as 1930 that shipping was not an end in itself, but rather a part of the worldwide ebb and flow of people and commodities. He wrote that shipping could not flourish when trade was sick, and he identified the main cause of the Great Depression as fear, not just among individuals, but also as it affected nations. He was surprisingly prescient in correctly identifying some of the fault lines that would fracture the fragile world peace by the end of the decade: 'Spain is concerned with revolution; France nervous about Germany; Germany afraid of Russia; Russia afraid of her own people; England of economic troubles; even America is perhaps afraid of these ... the result is slow paralysis of international trade, with corresponding evils for shipping.'[5]

Nationalism and politics had become inextricably bound up with each other during the early 1930s. The post-war rivalry between international shipping companies had intensified, and each 'great ship' that was launched was imbued with ideas of patriotism and aspirations to maritime superiority. The international shipping companies were sensitive to political developments in other countries, especially the ominous rise of the National Socialist Party in Germany. The issue of whether foreign visitors should offer the 'Heil Hitler' salute when visiting the country was broached in November 1933, just ten months after Hitler came to power, in the *Canadian-Pacific Gazette*, the on-board newspaper of the *Empress of Britain*. The leading German newspaper, the *Deutsche Allgemeine Zeitung*, had already recommended that foreign visitors should wear special identification badges, as 'Germany is anxious to avoid any further unpleasantness such as that which followed the attacks on British and American citizens who failed to give the Hitler salute ... Failing the introduction of the badge system the Deutsche Allgemeine Zeitung advises foreigners either to give the Hitler salute for the sake of peace, or else avoid all occasions in which it might be required.'

With Germany resurgent under Hitler, the rise of National Socialism had profound repercussions for that nation's merchant navy. Hitler intended to reward those supporters who had swept the Nazis to power in 1933. The 'strength through joy' initiative encompassed the building of two cruise ships, with single-class accommodation on board for model workers. The *Robert Ley* sported a swastika on its funnel, and was launched in 1938. Newsreel footage survives showing Adolf Hitler visiting the ship amid cheering crowds, who were delighted to be able to join the 'cruising classes'.

Needless to say, the limited opportunities for women hoping for careers at sea in German-owned ships rapidly diminished under the Nazi regime. In 1937 the *Daily Telegraph* reported that:

> Great indignation has been aroused in conservative German shipping circles by the news that a woman, Fraulein Annaliese Sparbier, formerly a schoolmistress, is working for a master's certificate, with a view to becoming a captain in the German mercantile marine. Fraulein Sparbier is now serving on board a trawler as an ordinary seaman. The 'Deutsche Seeman', the official organ of the German merchant service, asks severely whether such activities will help Fraulein Sparbier to 'do her duty as a woman and bring healthy and strong children into the world'. Her ambitions, it adds, are incompatible with the ideas of womanly tenderness and sense of duty. 'When Fraulein Sparbier doffs her thick shoes and stockings and her oilskins, and enters the haven of marriage, she will be sorry to find she has lost some of her charm. Surely, if her love for the sea is so tremendous, she could become a stewardess in a liner.'[6]

In Britain many women who had avoided entering 'the haven of marriage' were now gainfully employed aboard the *Queen Mary*, including the resourceful Edith Sowerbutts, who had been recruited as a stewardess for the Cunard flagship in 1936. Edith was offered the job because she was very experienced, competent and, incidentally, fashionably slim, despite her enormous appetite, in the same way that air hostesses were required to be glamorous and *soignée* in the Jet Age two decades later. As the *Evening Standard* had reported, the *Queen Mary* employed an unparalleled number of women working on board: among the ship's company were swimming instructresses, hairdressers, a masseuse, nurses and female switchboard operators, and there was considerable prestige in working on the *Mary*, as it was known informally. The stewardesses were managed by a chief stewardess who in turn reported to the purser. Their working conditions were now far better, as they had a simple rota system, with more time off during the working day. In addition, after four or five return transatlantic trips, there was a month ashore for rest and recuperation, during which the seawoman could sign on for fifteen shillings a week on the dole. There was always a risk that any resting stewardess might be 'crimped', that is posted to a lesser ship of the same line, if the company was unexpectedly short of personnel, but generally the *Queen Mary* maintained the same staff, and morale was high. Edith was always relieved to get back on board her favourite ship, as it carried the most prestigious and wealthy passengers, and the tips were excellent, as much as £20 per month.

Apart from the launch of the *Queen Mary*, 1936 was a tumultuous year. With the death of King George V in January, the Prince of Wales became Edward VIII, a prospect he viewed with dread. After a suitable period of court and public mourning, preparations began for his coronation, which was planned for May 1937. However, the illicit relationship between the prince and Mrs Wallis Simpson was already known to those in society and government circles. Within months the impasse would result in the constitutional crisis of the abdication. Meanwhile, the British press maintained a stout code of silence, though British subscribers to American journals were puzzled as to why large rectangles had been cut out of their magazines by the scissors of the censors. That momentous year ended with the abdication of Edward VIII, who rather melodramatically broadcast to the nation that he could not continue as king 'without the support of the woman I love'. Most of the British public were completely nonplussed; they had never heard of Mrs Simpson, but the story erupted in the press on both sides of the Atlantic and the American journalist H.L. Mencken described the romance as 'the greatest story since the Resurrection'.[7] Winston Churchill, himself the product of a marriage between an American *femme fatale* and a British aristocrat, wondered out loud why the king should not be allowed to have his 'cutie'. Noël Coward, acerbic as ever, replied, 'Because the British people will not stand for a Queen Cutie.'[8]

Wallis's countrywoman, Lady Astor MP, was in New York with her long-suffering maid, Rose Harrison, when the news of the abdication broke. Nancy was livid; as an eminent society hostess, she had long been aware of the romantic relationship between the prince and Mrs Simpson, and had tried to persuade the future king against continuing it. But Rose found her mistress's unsympathetic attitude difficult to understand. Both Wallis Simpson and Nancy Astor were ambitious American-born women, who had reinvented themselves by entering British society, finding acceptance in elite drawing rooms and royal circles. Both women had undergone the considerable stigma of divorce, having previously been married to volatile and violent drunks, before finding more gentlemanly second husbands through their travels across the Atlantic. But Nancy Astor was vehement that Mrs Simpson should not become queen, and cried bitterly when told what the New York paperboys were shouting below in the street as the news first broke. Perhaps she was concerned that the British public might now see her too as an 'upstart' divorced American woman. On her return to Britain she was asked to make a radio broadcast to the United States and she was at pains to explain that the abdication occurred because Wallis could not marry the future king as she was a divorcee, and not because she was American.

Rose speculated that it was some consolation later to Lady Astor that her great friends, the Duke and Duchess of York, were to accede to the British throne in place of Edward VIII. The coronation was still planned for May 1937, even though the *dramatis personae* had changed. Cunard embarked on a major promotional initiative to bring thousands of Americans to London to attend the celebrations, and, perhaps wisely, their marketing literature featured an impressionistic scene of the famous gold coach pulled by horses, with a tactfully indeterminate (though presumably regal) figure inside.

Meanwhile, Cunard pressed ahead with another giant ship for the transatlantic route. The new vessel was intended to be a replacement for the venerable *Aquitania*, which was due for decommissioning in 1940. Hull 552 was laid in December 1936 at John Brown's, and, following the abdication, it was given the name of the *Queen Elizabeth*, the new title of the Duchess of York. Queen Elizabeth had been furious when her brother-in-law abdicated as she had never intended to undertake the very public life expected of the consort of the monarch. When Elizabeth had finally agreed to marry Bertie in 1923, after turning him down twice, it had been assumed the couple would lead mostly private lives – a necessity given his shyness and bad stammer. Nevertheless, she took to the new role with as much grace as she could muster. At the height of the Munich Crisis at the end of September 1938, Queen Elizabeth travelled to Clydeside with her two young daughters to launch her namesake, with 300,000 people watching the ceremony to dedicate the world's largest passenger ship. The queen spoke of 'the great ships that ply to and fro across the Atlantic, like shuttles in a mighty loom, weaving a fabric of friendship and understanding'.

Such a benign, constructive image was conjured up despite the darker and starker global realities of the late 1930s. The inexorable rise of the dictators in the Old World – of Mussolini, Stalin, Franco and Hitler – was creating vast ripples across Europe. Those with foresight and overseas connections assessed their options, and some decided to leave their homelands by crossing the Atlantic. One woman who made her escape was the film actress and inventor Hedy Lamarr, whose real name was Hedwig Eva Maria Kiesler. She was born in 1914 in Vienna to a wealthy and cultured family of assimilated Jews. Her flourishing career as an actress attracted notoriety in 1933, when her fifth film, *Extase*, was criticised for its nudity and frank sexual content. When Hitler came to power in neighbouring Germany in 1933, the film was banned on the grounds of public morality. Hedy resumed her stage career, and the following year, when she was nineteen, she married an Austrian armaments manufacturer, Fritz Mandl, fourteen years her senior, also of Jewish ancestry. He tried to buy up all surviving copies of *Extase* in order to destroy them, though this merely inflated their price on the black market. Mandl was obsessively jealous about Hedy, his beautiful and elegant spouse, and constantly imagined that she might have an affair with another man. He insisted that they live in a grand but remote country house where he could entertain political and military contacts such as Mussolini. Hitler, however, would have nothing to do with Mandl because of his Jewish origins.

In 1937, months before Austria became part of the Third Reich, Hedy escaped. Her marriage to Mandl had become odious to her, and the political situation in Austria and neighbouring Germany was increasingly threatening people of Jewish origin. Those Jews and anti-Nazi activists who could were leaving in droves, and many creative people were heading to America, seeking work in Hollywood, including Billy Wilder and Fritz Lang. Hedy longed to resume her career as an actress; she loathed her husband's political machinations, his pro-Nazi contemporaries, and she disliked being a trophy wife, required only to look beautiful and say nothing, 'standing still and looking stupid', as she put it.

The various accounts of how Hedy managed to escape her husband's surveillance read like the plots of 1930s thrillers. One story is that she drugged one of the housemaids who resembled her, and while the servant slumbered in her bed, Hedy dressed in the girl's uniform and escaped on the maid's bicycle. She reached a station and took the train, eventually reaching Paris where she filed for divorce, before slipping across the Channel to London, where Mandl's influence was less strong. In truth, following a furious argument about her wish to return to the stage, her husband stormed out to spend the night in one of his hunting lodges. Before he could return, Hedy quickly packed some clothes and furs, and her jewellery, though she had very little cash. By her own account, 'I managed to leave Vienna that night, veiled and incognito and with all the trappings of a melodrama mystery. And I went straight through to London.'[9]

At a small party in London, she was introduced to the head of MGM Studios, Louis B. Mayer. He was scouting for new talent, recruiting talented actors who were keen to leave Europe because of the political situation, and naturally he knew the banned film *Extase*. He was impressed by Hedy, but only offered her a contract worth $125 a week, and that on condition that she paid her own way to America. She suavely declined his offer, but devised a plan to obtain a better deal. Mayer and his wife had already reserved their return passage to the States on 25 September on the elegant French ship, the *Normandie*. Hedy booked one of the more affordable cabins on the same voyage. Wearing a succession of gorgeous evening gowns, and her best jewellery, night after night she made the *grande descente* down the mirrored staircase into the first-class dining room, accompanied by a succession of wealthy and ardent young men. The glamorous setting of the *Normandie* showed her beauty to its best effect, and jaws dropped at her elegance, deportment and evident 'star quality'. Douglas Fairbanks Jr, a fellow passenger, accomplished actor and an important player in the film industry in his own right, could not take his eyes off her. As a result, Mayer scrambled to sign her up, this time offering her a starting contract of $500 a week if she could master English, and was willing to change her surname to something less Teutonic and more euphonious.

This voyage and her decisive action changed Hedy's life. She escaped her marriage, her old life and the repressive atmosphere of Austria. She had staked everything she possessed on a single transatlantic ticket, in order to try to secure a film contract. By the time she arrived in New York, aged just twenty-two, she had a new name, Hedy Lamarr, and was hailed as MGM's latest discovery, greeted by a barrage of press interest and flashbulbs. By October 1937 she was living in Hollywood; she starred in a film, *Algiers*, with Charles Boyer which was a huge hit in 1938. She was courted by John F. Kennedy, among others, and became friendly with the maverick millionaire Howard Hughes, but she married a much older screenwriter, Gene Markey. Her film career took off spectacularly, though her marriage foundered in the early 1940s.

It seems fitting that the splendid setting of the *Normandie*, with its mirrors, uplighting and grandstanding opportunities enabled Hedy Lamarr to secure a lucrative film contract. She exploited the dramatic potential of the *grande descente*, just as a consummate actress would time and judge her entry on stage. By the mid-1930s the ocean liner had become a potent symbol of the glamour, modernism and chic lifestyle perpetuated by the blockbuster movies of Hollywood. The theatrical interior design of the vessel appealed to the international set and the 'beautiful people' of Hollywood, influencing the set design of movies. Indeed, there were even movies set on ocean liners: the 1937 musical comedy *Shall We Dance* stars Fred Astaire and Ginger Rogers, who become romantically involved on an ocean liner on its way to New York. In a 'mixed race' dance number considered daring for its time, Fred Astaire encounters a group of African-American crew apparently holding a jam session in a spotlessly clean, art deco-style ship's engine room, and tap dances to their music, with rhythms inspired by the vessel's engines.

There was a real symbiosis between Hollywood, the heart of the movie industry, which acted as a 'dream factory' in the inter-war years, and the ocean liner, which on every voyage carried people full of hopes and aspirations. Not only did actors and actresses regularly travel the Atlantic to appear in films and stage shows, but there were also movie moguls seeking lucrative deals, theatrical impresarios and talent scouts who were travelling to other countries to find the 'next big thing'. Naturally, the publicity departments of each shipping line were glad to supply flattering photos and benign press stories about celebrities and royalty, free of charge, to newspapers and magazines, to persuade less well-known mortals to travel on their vessels in future.

Wise stars knew how the PR machine worked and played along with it; Mae West was considered a 'good sport' because she would willingly pose for press and in-house photographers. Marlene Dietrich was another much travelled celebrity during the 1930s and was frequently photographed and filmed on board ship. Glamorous and elegantly draped in furs, she was willing to co-operate with the press, seeing her public persona as part of her job. A thorough professional, while on board ship she never appeared in public for breakfast, rarely at lunchtime, but would make a spectacular entrance at dinner. Her favourite table in the *Queen Mary* dining room was the most prominent, the one preferred by her friend Noël Coward, but fortunately they never travelled on that ship at the same time.

Edith Sowerbutts met many celebrities including Robert Taylor, David Niven, Gary Cooper, Doris Duke and Douglas Fairbanks Snr. She described Paul Robeson as 'every inch a gentleman', and he was a regular on the *Queen Mary*. He remembered the names of the crew, and often mingled with them in the Pig & Whistle, the crew's after-hours bar, even performing there during one voyage. Edith recalled his beautiful speaking voice, and she regretted missing the chance to hear him sing, as female crew members were not allowed to join the men in the Pig & Whistle.

For the women who worked on board the great ships, proximity to glamour was one of the appeals of the job. In an era when film stars and actors were international household names, there was considerable cachet to being intimately involved in caring for the requirements of these stellar figures when they travelled by ship. Of course, despite their glittering careers and constant appearances on the silver screen, the ocean-going famous were mere mortals, constantly ringing for a hangover cure, breakfast in bed, a massage in their cabin, an attractive snack on a tray, advice on avoiding seasickness, the marshalling of their entourage, and the endless bringing and despatching of messages, flowers and gifts.

Celebrities had their own favourite ships. Film star James Stewart, French-born novelist Colette and Josephine Baker loved the *Normandie*. The French Line's *Île de France* was the favourite of Gloria Swanson, Yehudi Menuhin and Arturo Toscanini. Ella Fitzgerald and Duke Ellington preferred the *Queen Mary*, which Cary Grant called the Eighth Wonder of the World. Bing Crosby was also a regular traveller on the *Queen Mary* and he was a keen photographer. He became friendly with the on-board Ocean Pictures photographers and would often join them in the darkroom as they developed that day's crop of images.

Until 1939 Edith Sowerbutts worked on the *Queen Mary* as a stewardess. The ship was a great improvement on her previous vessels: there were service lifts between the kitchens and the small deck pantries where each stewardess was based, and a phone linking the two so that passengers' food orders could be relayed to the kitchen clerk. Edith noted that the ship's designers had obviously never tried to lay a silver service tray for breakfast in the tiny pantry, but otherwise the labour-saving devices were appreciated.

Edith's own breakfasts on the *Queen Mary* were brought to her by the deck waiter. She usually had orange juice, coffee, bacon and eggs, toast and marmalade – she had a big appetite. She consumed it standing up, between responding to early morning passenger calls, taking breakfast trays to her passengers, and bedmaking and dusting her cabins once her passengers were safely out of the way. At mealtimes Edith and her shipmates would often find an unoccupied cabin where they could eat a snack, and they would meet again later for a clandestine cocktail and cigarette before starting their evening's work, which involved tidying staterooms, putting away clothes, and turning back bedcovers while the passengers were dining or dancing. Their own supper would be between 8 and 9 p.m. The quality of the food was excellent but stewardesses rarely had a chance to eat a whole meal undisturbed.

Until 1939 stewards and stewardesses on the *Queen Mary* worked every day while at sea from 7 a.m. until 10 p.m., and would usually have two hours off in the afternoon. This was an opportunity for the women seafarers to mingle socially; they might sit in deckchairs on a small closed-off section of the crew's deck, with a packet of cigarettes, some knitting and plenty of ripe gossip. There was occasional friction: some were former domestic servants, such as housemaids, parlour maids, or ladies' maids, and those who had previously worked in grand households were sometimes suspected of suffering from 'folie de grandeur'.

Accommodation for female crew was still very cramped. Edith shared a tiny cabin with her old friend Ada Norfolk, another stewardess. On getting up at six thirty every morning, they had to take it in turns to dress, as there was limited floor-space. Edith would wait on the top bunk while Ada ritually donned her stockings and shoes. She would leave the stockings concertinaed around her ankles ('Russian boots, dear,' she remarked wryly) until she could get her suspender belt on and attach the stockings. Underwear came next, and eventually she clambered into her grey uniform dress and attached her white cap to her hair. Only once Ada was dressed and tucked up neatly on the lower bunk was there enough room for Edith to dress. They set out on a five-minute walk from their cabin to their stations at midships.

Ada Norfolk had previously worked as a senior stewardess on the *Olympic*, *Majestic* and *Berengaria*. Perhaps uniquely, she had run away from the circus, and ended up going to sea. She came from a long line of performers and showmen; her father was a clown and had his own circus, and her brother was a juggler. Many years before, the family were sailing to America to fulfil a circus booking, when the purser took a shine to young Ada. Despite their eighteen-year age difference, they married. When he died, Ada was left to bring up two young children. As a 'company widow', she was employed as a stewardess on United States Lines, where she had met her husband. Ada had survived the torpedoing and sinking of a ship off the coast of Ireland during the Great War, and described the experience in a very matter-of-fact manner: 'It was a lovely day in June, dear. It was quite pleasant in the lifeboat for a while, and then we got picked up.'[10] In idle moments, Ada fondly recalled that in the early years of her marriage, during a brief sojourn in California, she had met a pleasant and handsome young British chap, anxious to make his mark in Hollywood. This unknown was called Charlie Chaplin.

The chief stewardess of the *Queen Mary* was Mrs Nin Kilburn, who always looked after travelling royalty. Mrs Kilburn came from a Liverpool sea-going family – one of her sisters and an aunt were stewardesses on the *Lusitania*. When that ship was torpedoed in 1915, a deck steward tried to save the sister, but she insisted on staying on board until she found her aunt. In the confusion, the aunt was saved, but Mrs Kilburn's sister drowned. Nin Kilburn was a former school teacher; in the 1930s, when unemployment was high, it was not unusual for well-educated and well-qualified women to pursue better-paid careers at sea. A talented linguist, she was paid an extra £1 a month for interpreting French or German.

While long-lasting and valuable relationships often developed between women seafarers who shared adversity, seasickness and tiny cabins, there were also friendships between them and some of their passengers. Some proved to be superficial and illusory: Violet Jessop recalled one well-heeled American socialite who was recovering from a personal calamity, and who relied on her emotionally throughout an entire transatlantic voyage. They spent many of Violet's few and precious off-duty hours in 'intimate and soul-revealing talks'. On arrival in New York, the grateful passenger insisted Violet must come and visit her any time she was in the city. A few months later, Violet dropped in at the plush hotel where her acquaintance lived. She quickly realised that the woman seemed puzzled as to how they knew each other. She was welcoming, but Violet recalled, 'I knew she had not the faintest idea who I was.'[11]

Some wealthy travellers would book crossings on specific ships in order to be in the care of their favourite stewardesses; in fact Cunard would offer the services of a named 'special' stewardess at a supplementary charge for those passengers prepared to pay extra for a familiar face. Miss Paddock, who spent her working life as a stewardess with White Star Line, always looked after Harriet Cohen, the professional pianist, on her transatlantic trips. They became great friends, and Miss Paddock was one of the guests on *This is Your Life*, a popular British TV programme, when it featured the musician's biography. Another famous pianist, Dame Myra Hess, was usually attended by Janet Austin on the *Queen Mary*. When Miss Austin reached the age of sixty-five and retired,

Dame Myra provided her with a small flat at her home in St John's Wood in London, to ensure their friendship continued. Many stewardesses would be sent money at Christmas by some of their former passengers, usually in the form of $20 bills, and some were even remembered generously in their wills. Ada Norfolk, Edith's cabin mate, unexpectedly inherited a substantial legacy long after she retired, left to her by a former Tiller Girl with whom she had been friends when they were both young and had crossed the Atlantic together between the wars.

Stewardesses were careful not to leave valuables or money in their own cabins, especially when in port, in case of theft. By the end of a return voyage they had often accumulated a substantial sum in tips from grateful passengers. This could be stored in the purser's safe, or they could deposit it with the branch of the Midland Bank on the *Queen Mary*. Traditionalists, however, preferred to conceal it on their persons, either in a special reinforced pocket sewn into an old-fashioned petticoat, or in a purse secured around the midriff, and some stewardesses developed mysterious lumps and bumps under their uniforms. As soon as they disembarked in Southampton and had been paid, they would visit the Post Office and bank their wages, before heading back to their families. Edith recalled: 'It was comforting for all of us, we female sea-dogs, on arrival at home, to be able to meet all the bills and commitments and no longer have to stint ourselves personally. My own responsibilities had always halved my earnings. Alone I could keep myself quite well. Now in the *Queen Mary* the money I made was a veritable godsend.'[12]

The stewardesses often had to deal with elderly ladies losing their false teeth. Passengers were very partial to removing their uncomfortable dentures to consume soft fruit in the privacy of their cabins, and they often forgot to re-insert them after they had finished eating. If they were unlucky, a dish of discarded peel, pith and cores might be flung through the open porthole, without the stewardess spotting a set of dentures on the same plate. Passengers were also adept at purloining the silver cutlery, as well as the distinctive silver cruet sets, much valued as souvenirs. A stewardess's ruse for retrieving them was to insist that she needed them 'for cleaning'. Eventually the cutlery and cruet sets were stocked as merchandise for sale in the on-board ship, along with the attractive white cube-shaped teapots and milk jugs designed specifically for the Cunard Line. Passengers were encouraged to buy the tea sets as souvenirs, rather than pilfering them from their breakfast trays.

The single aspect of life on board the *Queen Mary* that Edith most feared was a storm at sea. Despite rarely suffering from seasickness herself, she would listen with apprehension to the warning signs as a storm approached. There might be a clatter of metal trays or crockery from the stewards' pantries as the ship lurched or rolled. Baggage would start to slide across the floors of the cabins, and would have to be secured by the stewards. Carpet would be placed on top of polished floors to enhance grip, while stewards and waiters in the public rooms would secure all the chairs to pillars or walls, using lanyards so that they couldn't slide about. However, so dramatic could the force of a storm be that occupied chairs had been known to break away in the dining saloon, careering from one side of the room to the other, taking their passengers with them.

The *Queen Mary* was notorious for its lively performance in heavy seas, a fact that Cunard was keen to conceal from the travelling public. Smaller ships would sail up a wave, crest it, and hurtle down the other side, bobbing on the water like a cork, and providing a sensation for the passengers like being on a roller-coaster ride. But the vast *Queen Mary* – nearly a quarter of a mile long – would labour up a wave and seesaw on its apex before plunging forward and downwards, a much more vertiginous drop for anyone travelling in its stern or bows. It was estimated that the difference between the crest and the trough of a wave during a severe storm could be the equivalent of eight storeys in height. The ship was also prone to rolling from side to side, and a list of more than 40 per cent was recorded on a number of occasions. During one ferocious storm in the late 1930s that lasted for five days, one stewardess rolled right up the bulkhead of her cabin and back again. She remembered never having seen so much damaged crockery and china, the alleyways were running with seawater, and even some portholes had been smashed by the force of the waves. While beleaguered and frightened passengers were wedged into their bunks with pillows to keep them both prone and safe, the cabin crew – struggling to stay upright themselves as the ship pitched and lurched – attempted to minister to them, bringing dry biscuits and bottles of Canada Dry ginger ale in vain attempts to combat seasickness.

A particularly memorable storm at sea occurred in April 1938. The *Queen Mary* finally reached Plymouth five hours later than expected, after experiencing a terrifying ordeal. An eighty-mile-an-hour gale had driven the ship eastwards, with waves more than a hundred feet high crashing over the decks. As the storm reached its crescendo Miss Lily Pons, a professional soprano singer, defied the howling winds and the suspension of the normal rules of gravity to star in a public concert for charity. The 'pocket prima donna', as she was known, was determined that the show should go on, though Miss Pons had had an inkling that she might be in for a salutary experience:

> The night before, my bed crashed into the stateroom wall. My trunks and all the furniture in the room were piled up in a heap. Stewards came and clamped down my bed. But I was determined to keep my promise. The ship was rolling so badly when the concert began that safety ropes were placed in the room, so that the audience could hold onto them. When I started to sing, I was clutching a rope. I let go of the rope unconsciously – and the next thing I knew was that I was sliding along the stage. I could hardly keep my feet but I went on singing, and £300 was collected for seamen's charities.[13]

The gale lasted twenty-four hours and was physically and mentally exhausting. Passengers and crew were repeatedly tipped off their feet, furniture was smashed, and the damage to crockery and glassware was unprecedented. A grand piano broke loose from its moorings in the salon, and swept across the room like a three-legged behemoth on castors – fortunately, people in the vicinity managed to leap out of the way. An American banker who was dozing in a reclining chair when the ship suddenly lurched sideways, woke up in the scuppers with a broken arm and a black eye. A theatrical producer, Marc Connelly, said: 'It was the worst storm I have ever experienced, and I have crossed the Atlantic thirty times. Fortunately, the gale was behind us but the seas were like a mountain range of water. I have never seen anything so awe-inspiring.'[14] Mr Connelly said he saw a dozen people being carried away by the crew and brave passengers for medical attention. It took six stewards to lift one man, who was unconscious. Ray Noble, the *Queen Mary*'s dance band leader, said he was surprised there were so few casualties. When the *Queen Mary* put in at Plymouth, forty injured passengers were taken off to be treated for injuries in local hospitals, and the ship limped back to Southampton for substantial internal repairs and provisioning.

But 1938 was a memorable year for tumultuous events on land as well as at sea. Tensions in Europe had been building steadily as the true nature of Germany's Third Reich became clearer, to its own nationals, its neighbours and perceptive international observers. In particular, for those who worked in transatlantic travel, or who travelled regularly, it was apparent that many more Europeans were on the move. They were not travelling abroad for leisure, pleasure, romance or business: they were running to escape from what appeared to be the growing menace of anti-Semitism. Discrimination against the Jews had occurred in cyclical waves throughout Europe since the Middle Ages, but during the 1930s the economic consequences of the Great Depression led to increasing animosity against them. Some politicians looked for scapegoats to account for their national malaise, and the Jews were convenient and easy to blame. France had a number of very active right-wing political factions, and in Britain Sir Oswald Mosley founded the British Union of Fascists, ostensibly as a force against unemployment. Before long, his organisation was involved in street battles with recent Jewish immigrants living in the East End of London. However, it was Germany under the National Socialists – and, as the decade wore on, its annexed or overrun neighbours – whose Jewish communities were most desperate to escape by emigrating, ideally to North or South America.

As the conditions in Europe became more hostile to the

Jews, the shipping companies responded. As early as November 1933, just ten months after Hitler came to power, the need to increase the capacity of the kosher kitchen on the *Aquitania* was discussed by the Cunard board. It was apparent that increasing numbers of Jewish emigrants were planning to leave Europe, especially those of German origin. Understandably, they preferred not to sail on German-owned ships, preferring French, American or British shipping lines where they were treated more respectfully. Kosher food was provided and listed on the menus of all Cunard passenger ships, and separate kitchens were maintained to comply with strict dietary rules. Provision had also long been made on the more enlightened ocean liners for religious observance for Jewish passengers. *White Star Magazine* noted that:

> for Jewish ocean travellers crossing the Atlantic in the latter part of April [1927] special arrangements were made by the White Star line. During the voyage of the *Olympic* which began at Southampton on April 15th, Passover services were held, three rabbis who were travelling in the ship having charge of these. Special steps were taken to ensure the proper preparation of food in accordance with the Jewish ritual.[15]

The *Queen Mary* was the first ocean liner to be equipped with its own Jewish prayer room, and Cunard appointed a rabbi to ensure a kosher kitchen and catering department for observant Jewish passengers. The numbers of those planning to leave Europe grew rapidly in the 1930s, though total immigration numbers to the United States for each nationality were still limited by the 1924 quota laws. For many, the great ships provided a lifeline; if they could obtain a visa and raise the funds to buy the tickets, they could sail for the New

World. By 1938 about 150,000 German Jews had already fled the country, and they were now spread across the globe; some had found sanctuary as far away as Shanghai. Following the Anschluss, the annexation of Austria in spring 1938, a further 185,000 Jews found themselves living under Nazi rule.

As described in the *Washington* Post, two families' tales illustrate the dilemma. Austrian-born businessman Mark Tennenbaum had anticipated the growing menace and had planned an escape route, sending money to a trusted friend who was living in neutral Switzerland. Through his contacts, the friend was able to arrange American visas for Mark and his wife, Earnestine, and their two-year-old son, Robert, and as instructed he also bought first-class tickets for the three of them to travel on the *Queen Mary*. Mark Tennenbaum had reasoned, 'Give the money to the Brits, not the damn Nazis!' because punitive measures imposed on emigrating Jews by the Nazi regime included confiscating all their money as they left the country, leaving them only the equivalent of $4 per person in ready cash. The Tennenbaums visited their friend, ostensibly for a holiday, then travelled from Switzerland to Cherbourg to board the *Queen Mary*, clutching their precious visas and tickets.

Mark was a keen amateur cameraman, and film survives of the small family exploring the deck of the *Queen Mary* as they sailed to America and a new life. The footage shows an excited small boy, Robert, in a double-breasted overcoat, short trousers and round sunglasses, holding the hand of his elegant mother, who is wrapped in a smart outfit with a fur collar and hat. She is pointing out the features of the ship, and encouraging him to make friends with another youngster. But in an unguarded moment, for a few seconds Earnestine gazes into the camera lens; her smile has been replaced by an expression of apprehension and regret. 'She looks so sad, and it was for a good reason,' Robert commented, viewing the film again, eighty years later. The Tennenbaums had experienced increasing anti-Semitism, and had been obliged to abandon two thriving businesses, their property and possessions in order to make their escape. Nevertheless, he recalled in old age, 'The bottom line was that the *Queen Mary* saved me and my mom and dad, saved our lives.'[16]

The noose tightened on later émigrés: three weeks before Kristallnacht on 9–10 November 1938, the night when thousands of Jewish businesses and properties were attacked and destroyed, Ludwig Katzenstein's father resolved it was time for the family to escape their home near Berlin. He had secured emigration visas for the family, and using their dwindling fund of Reichsmarks he purchased four last-minute boat tickets on the *Queen Mary*, which was leaving from Cherbourg for New York. The family took the train through Germany, but at the border with Holland the train was stopped, and all the passengers' papers were checked by the Gestapo. A new edict had been introduced just the day before their flight: all Jewish citizens were required by law to have a red letter 'J' stamped into their passports to make them valid. The Katzensteins' documents were not in order because they lacked this new stamp. The family were detained and put into a holding cell, while the train left without them. Ludwig was only six years old; he remembered his father was allowed to go into town, probably to pawn something to get enough money to bribe the guards, and on his return the letter 'J' was added to their passports and the family were free to leave on the next train. However, it was now evening, they were now many hours behind schedule, and in danger of missing the ship. Ludwig's resourceful father persuaded the train master to call the captain of the *Queen Mary*, to ask the ship to wait. He knew that if they missed this sailing, they would have lost their last chance to get to the USA, as they had no more funds. Commodore Robert Irving received the message and delayed the departure of the great ship and its thousands of passengers and crew by six hours, so that the Katzensteins could scramble up the gangplank and on to the safety of the *Queen Mary*. Ludwig recalled: 'My father asked him to wait until four people got there. That captain waited for six hours. He saved our lives ... it was a miracle. It shows, in my mind, considerable humanity.'

After the repression and mounting sense of fear they had experienced in Germany, the voyage on the *Queen Mary* was a revelation. 'We ate kosher on there, and they had a synagogue. We had services in that synagogue, and on Shabbat they had special services. It was just wonderful.' When they reached New York, and saw the Statue of Liberty, he remembered, 'You felt free for the first time, after so many years.'[17]

Those working on board the transatlantic ships were aware of the palpable fear among most of their Jewish passengers as the international political situation deteriorated. Edith Sowerbutts, while working as a stewardess on the *Queen Mary* in 1938, was located on B deck for one voyage, and the cabins in her care were en route to the ship's Jewish prayer room, which was open to all classes of passenger. Every day, she would encounter Orthodox Jews from third class nervously asking for directions to the synagogue, and she and the male bedroom steward would direct them. Among the first-class staterooms on B deck were some cabins occupied by a cosmopolitan group of Dutch Jews. They were obviously accustomed to wealth, but they had left everything behind them, convinced they knew what would happen if they stayed. At the end of the voyage, Edith and the bedroom steward were summoned together to the staterooms, to be thanked with courtesy, and offered a substantial tip each, which she was sure they could probably ill afford. The steward, whom Edith had previously considered a rough diamond, politely declined, saying: 'From people like you, we take nothing. We thank you, the stewardess and I, and wish you luck. It has been our pleasure to look after you.'[18] Edith wished she had thought of that speech.

Dorothy Scobie, who came from a working-class Liverpudlian family, was enraptured by the sight of the models of ships in the Cunard Building when she was a child, and longed to go to sea. She was employed as a stewardess on a number of the less prestigious Cunard-White Star ships crossing the Atlantic from Liverpool, from 1937 till the outbreak of war, and her career afloat eventually spanned more than twenty-three years. Fares on the Liverpool to New York route were usually keenly priced because the ships were smaller and older, but Dorothy noticed there was a rising demand from European passengers needing the cheapest possible one-way tickets to America. In the summer of 1939 she also observed a psychological change in her third-class passengers. Many of them were refugees from Russia, Hungary, Germany, Latvia and Austria, who had already made arduous journeys from their homes to get to a British port, and across country by rail to Liverpool to pick up a ship to the United States. They were perpetually anxious and seemed poor; in particular, many of the Jews fleeing Germany and Austria by this stage had very few possessions, as they had either escaped in a hurry with what they could carry, or had had all their property confiscated by the authorities before they left the country. Dorothy recalled:

> These people had embarked in Liverpool and none of them had big trunks. Always they counted and recounted their dozens of suitcases and brown paper parcels. Hat boxes, string bags and attaché cases. Most of the men had briefcases which they never let out of their arms ... What sorrows had they left behind? Indeed, whom had they had to leave behind? What traumas had they already witnessed in their short lives?[19]

Many European refugees followed on the transatlantic ships, hoping to escape the coming conflict, but not all would find sanctuary.

9

Women under Fire

On the morning of Sunday, 3 September 1939, wireless sets all over Britain were turned on and tuned in to the BBC Home Service. The nation waited with bated breath while a refined female voice finished dictating a recipe for shepherd's pie. After a short pause, at eleven fifteen the Prime Minister, Neville Chamberlain, veteran of the Munich Agreement, took to the airwaves. He made the solemn announcement that as the British government had received no response to their final ultimatum to Berlin, the deadline had expired, 'and consequently, this country is at war with Germany'.

Although the outbreak of hostilities had long been anticipated, the actual declaration of war came as a shock. All over the globe there were people and goods in transit. Previous plans were abandoned, and new ones made, as expatriates of all nations headed for home, or hurried to leave one country for another.

The *Queen Mary* was at sea when war was declared; it had left Southampton on Wednesday, 30 August 1939, heading for New York via Cherbourg. This was to be its last voyage as a passenger ship for six years, and every berth was taken. Public rooms had been turned into dormitories, and the capacious baggage alcoves on the main decks were converted into berths for six passengers at a time, with curtains hastily installed for privacy. There were Americans aboard who had signed on as crew, working their passage to get back to the USA. They were employed in the kitchens, washing dishes or chopping vegetables. Many of them returned to Europe within a few years to fight in the forces helping to free Europe, travelling once again on the *Queen Mary*, now a troopship. On board there were a total of 2,331 passengers and a crew of 1,231. Among the stewardesses were Edith Sowerbutts and Nin Kilburn. On the passenger list was the world-renowned theoretical physicist Albert Einstein and his wife Elsa, and the comic movie actor and radio personality Bob Hope, with his wife. When the news reached the ship that Britain had declared war on Germany, Bob Hope performed a special show for the passengers, singing his signature tune, 'Thanks for the Memory' with rewritten lyrics.

The same day that war was declared, the first British merchant seamen and women were killed in the new conflict. At 7.45 p.m., less than nine hours after Chamberlain's broadcast, a German submarine torpedoed a British-owned Donaldson Line ship, the *Athenia*, 200 miles off the northwest coast of Ireland, before surfacing to rake the vessel with gunfire. Three-quarters of the passengers on board were women and children, many of them American. The *Athenia* sank and 118 died, including five stewardesses and fourteen male crew members. The attack on the *Athenia* was in direct contravention of the Anglo-German Agreement of 1935 restricting submarine warfare.

When the *Queen Mary* docked in New York on 4 September 1939 the ship was ordered to remain in port alongside the *Normandie* until further notice. After the passengers left, Edith and her colleagues hurriedly disembarked; they were being sent home immediately on another ship. A team of painters was already at work on the *Queen Mary*, painting its hull battleship grey. The *Mary* was left with a skeleton workforce to sail the ship back to Britain at a later date. Meanwhile, the rest of the crew were bussed downtown to join the *Georgic*, which was already loaded with passengers. 'This is the end of our lives, girls,'[1] predicted Nin Kilburn, the chief stewardess, anticipating unemployment (though in fact she was called back to serve on the *Franconia*, the ship that was to take Churchill to Yalta at the end of the war).

The *Georgic* had also been camouflaged with grey paint, and seemed small, cramped and shoddy after the vast and luxurious *Queen Mary*. Decks A and B were allocated to passengers, while female Cunard crew were housed on C and D decks, in tourist-class accommodation. Edith shared a four-berth cabin with just one other stewardess, a tippler who dosed her morning tea liberally with whisky. Edith objected to the fumes in an enclosed space at such an early hour. The ship zigzagged alone across the Atlantic, with no protecting convoy, a long and hazardous voyage, with the constant menace from U-boats. Lifeboat drills were frequent, and life jackets and gas masks were carried at all times, so great was the threat of imminent attack. Edith also took with her everywhere her 'ditty-bag' – a fabric carry-all containing whisky, bandages, cotton wool, safety pins, aspirins, 'plus the very necessary items for female hygiene. One never knew if all the excitement or stress might precipitate "the curse". Some among us were quickly inconvenienced by periods arriving too soon, doubtless due to the general upheaval of hurriedly changing ships, plus apprehension of war. Not me.'[2]

Edith also kept a tight grip on her money, the generous tips that she had garnered from her last *Queen Mary* voyage, in a purse safely secured to her suspender belt, along with a cherished and unopened bottle of perfume, a gift from an American passenger. This perfume was the last she would own for a number of years, and was called Froufrou of Gardenia, a floral scent she always associated with sailing days on the glamorous ocean liners.

The flight of the *Georgic* was a blend of adrenalin and boredom, as the seafaring professionals were unaccustomed to spending free time on board a ship. The refugee stewardesses had no specific duties, given that the ship already carried a full complement of staff, but out of habit and with few alternative clothes, they wore their uniforms every day, adorned with any jewellery they possessed, in case the ship was torpedoed. Mrs Kilburn, suddenly relieved of her normal working duties, learned to play bridge, and it became her abiding passion. She developed into an accomplished and formidable bridge player, and the game occupied her throughout her retirement, providing her with a circle of friends well into old age (she died at eighty-four). Edith also played cards with the men from the *Queen Mary*, the waiters, stewards and bellhops, to pass the time when they were not on submarine watch on deck. Certain somehow that there wasn't a 'torpedo with my name on it', Edith also took lengthy, luxurious early morning baths – this was reputedly a favourite time of day for submarine attacks, so it was a risk.

The possibility of a watery grave seemed very real to most people aboard. The church service held on Sunday morning was uncharacteristically well-attended, and the congregation sang 'Nearer my God to Thee', thought to have been the last hymn played by the *Titanic*'s orchestra in 1912 as that ship went down. The Cunard stewardesses clubbed together to buy a small bottle of whisky, costing six shillings, from the ship's bar, to be brought with them 'for medicinal purposes', if they had to take to the lifeboats. However, every evening, as darkness fell, bringing relative safety, the temptation to mark surviving another nerve-jangling day afloat proved too strong, and they would share out the whisky as a tipple to aid morale. Consequently, they would have to buy a replacement bottle the following morning.

The tension increased as the *Georgic* approached the Irish coast, favourite haunt of U-boats in the Great War. At night the ship was completely dark and no one was allowed to smoke on the outside decks. The ship finally sailed up the Thames to the London docks. The buildings along the shore were blacked out, a contrast with their last sight of land – vibrant, brilliantly lit New York. A disembodied voice came from a small craft floating below on the Thames: a naval officer speaking through a megaphone hailed them through the murk. 'Who are you?' '*Georgic*, from New York,' came the reply from the bridge. The ship docked, and the relieved Cunard staff and crew felt they deserved a celebratory drink before they disembarked. A waiter volunteered to go landside and bring back supplies. He returned with just three bottles of light ale; all he could find. Edith and the stewardesses realised they really were at war, and the next morning they scattered to their homes, in London, Edinburgh, Southampton and Liverpool, deflated and anxious. 'I admit to feeling slightly low spirited as the taxi deposited me at my garden gate one grey day late in September 1939. For over 2 years I had been a member of the crew of the liner Queen Mary. I now had the feeling that my life would never be the same again. It was not.'[3]

Both Edith and her sister Dorothy, who had been on the *Britannic*, managed to get home to the house they shared

with their mother. They were no longer required as stewardesses, because their ships had been requisitioned for war
service, so they reluctantly took clerical jobs on land. Edith
discovered that her membership of the National Union of
Seamen entitled her to 'danger money' for her hazardous
return journey across the Atlantic in wartime. She received
an extra sixteen shillings, the equivalent of the Sowerbutts
family's weekly grocery bill.

Meanwhile Edith's former ship, the *Queen Mary*, stripped
down, armed, camouflaged and fast enough to outrun submarines, became known as the Grey Ghost, shipping Allied troops
to theatres of war all over Europe and the Middle East. Hitler
offered a bounty of $250,000 and the Iron Cross to any U-boat
commander able to sink it, but without success, though the
Queen Mary had a particularly narrow escape in 1942. A
Nazi radio station erroneously announced that the ship,
packed with US troops, had been sunk off the coast of Brazil,
but in fact the plot had been foiled. A young American
diplomat based in Rio de Janeiro, John Hubner, had discovered that a suspiciously large radio transmitter had been
imported to Brazil by the German firm Siemens and Company,
and was being held for delivery. Hubner persuaded the
Brazilian police to mount twenty-four-hour surveillance on
the Siemens store. A German arrived to pick up the transmitter, was arrested and interrogated. He gave up the names
of his associates, and the location of a Nazi radio station in
the hills above Rio de Janeiro. Hubner and the police rounded
up the gang, and the radio station outside Rio was also
raided, just as it was transmitting a message to Nazi submarines regarding the sailing of the *Queen Mary*, which had
put in at Rio for fuel and supplies. The ship was far too big
to hide, and German spies had learned its top-secret sailing

time and its route and planned to transmit this to summon lurking U-boats. The discovery of the plot and the seizure of the radio station led to an immediate change of plan for the *Queen Mary* and it escaped before the U-boats could target the ship, but so certain was Berlin their attack would be successful that it prematurely announced the sinking. The former *Queen Mary* served for the duration of the war and would prove decisive during the D-Day invasion. Together with the other Cunard flagship, the *Queen Elizabeth*, the *Mary* transported more than a million troops as part of the war effort.

The *Queen Elizabeth*'s top-secret flight to America was one of the most audacious escapes of the conflict. The huge new Cunarder had still been under construction on the Clyde in 1939 when war was declared, and it was essential it was taken to America for completion and to escape possible damage from German bombers. The *Elizabeth* was hurriedly given a rudimentary fit-out, and painted grey, and by spring of 1940 it was generally anticipated that at some point it would undergo initial sea trials, but details were intentionally kept vague. On 3 March 1940 Hugh McAllister, the husband of Cunard's swimming star Hilda James, was on board the *Queen Elizabeth*, testing the newly installed radio equipment. He was puzzled to hear the engines were running, and went out on deck, only to discover that the ship was heading towards the sea. No warning had been given; the captain and the senior officers had received top-secret instructions and had sailed, regardless of the numerous workmen still on board. Even the captain had been left in the dark about the ship's true destination: he believed they were merely heading round the British mainland to Southampton, but once they had gained the open sea and he opened the second set of

Scottish teenager Mary Anne MacLeod left her home on the Isle of Lewis to sail from Glasgow to New York on the *Transylvania* in 1930. She had high hopes and $50 in her purse, and was seeking work as a domestic servant.

Popular entertainers were in great demand in the 1920s and 30s. Internationally renowned brother-and-sister dance act, Adele and Fred Astaire, frequently travelled on the 'Atlantic Ferry' for theatrical engagements in Britain and America.

Thelma Furness's affair with the Prince of Wales ended abruptly when he learned of her whirlwind romance with international playboy Aly Khan on board the *Normandie*.

Pampered passengers could spend their afternoons afloat in reclining teak steamer-chairs on deck, breathing in the invigorating ozone, with a cosy rug over the knees, while attentive deck stewards hovered with trayfuls of tea and cakes.

A Cunard White Star poster for the *Queen Mary*, 1936. The new flagship was designed to dwarf all other contenders on the lucrative and prestigious transatlantic route.

The first-class dining room of the *Queen Mary* could accommodate 800 passengers in a single sitting. The stylised wall map of the North Atlantic featured an illuminated crystal model of the ship, which moved every day to show the vessel's progress across the ocean.

The cabin observation lounge and cocktail bar of the *Queen Mary* was semi-circular in shape and had a curved zinc bar, with red and chrome uplighters. It was the most overtly Art Deco public room on board, and a popular spot for aperitifs and digestifs.

Two women passengers in a cabin on the *Queen Mary*.
Cunard's promotional material claimed that the rooms would
'convey the atmosphere of restfulness and comfort associated
with the most dignified British country homes.'

The *Queen Mary's* maiden voyage to New York, in 1936, culminated in
a rapturous welcome from a flotilla of other vessels. The quayside was
packed with cheering well-wishers, aeroplanes flew overhead in salute,
and fire hoses played over the Hudson River, casting rainbows in the
sunshine with their arcs of spray.

Marlene Dietrich appreciated the value of good publicity, and willingly posed for press photographers when travelling on ocean liners, following the advice of her friend Noël Coward: 'Always be seen, dear, always be seen.'

Austrian-born Hedwig Kiesler embarked on the *Normandie* in 1937, determined to persuade a fellow passenger, the Hollywood magnate Louis B. Mayer, that she had star quality. She arrived in New York a week later as MGM's latest signing, renamed Hedy Lamarr.

Maida Nixson reluctantly went to sea as a stewardess in 1937, and defied wartime torpedoes to escort evacuees to safety around the globe.

SPECIAL PICTURES OF CIVIL DEFENCE HEROISM

4th October, 1940

THE **WAR**

No. **50**

3D

WEEKLY

Incorporating WAR PICTORIAL

HITLER'S FOULEST DEED—THE MERCY SHIP MURDER

When an Atlantic liner carrying evacuee children to Canada was torpedoed by a U-boat in a gale, 600 miles from land, eighty-five out of 98 children lost their lives. The death roll totalled 306 out of the 421 passengers and crew aboard. Our artist's impression shows a passenger towing a raft—carrying adults and children—out of the suction of the heeling liner. The people on the raft and their rescuer were later picked up by a British warship after many hours at sea..

The War magazine reported the sinking of the British ship,
The City of Benares, in 1940. Public outrage was so great that the
transatlantic evacuation of children was abandoned.

Martha Gellhorn, inveterate war correspondent, was so determined to cover the D-Day landings in 1944 that she stowed away on board a hospital ship.

GI brides and their babies on the gangplank of the *SS Argentina* in Southampton in 1946. Operation Diaper Run reunited newly married British-born wives with their husbands in America.

sealed instructions, he discovered they were to head to New York immediately, at full speed, in order to outrun any U-boats. Hugh McAllister therefore sailed to America with the *Queen Elizabeth*; he had no choice. His work now was to complete the radio fitting aboard the ship in New York, and he was not able to return to Britain for nine months.

For many female seafarers, their chances of working at sea shrank rapidly during the early days of the Second World War. Passenger ships were requisitioned for troop transport, and hardly any civilians now chose to travel. But some passenger traffic persisted despite the hostilities, and brave seafaring women were willing to run the risks of being torpedoed or drowned. Maida Nixson, an English-born former journalist and writer who had fallen on hard times financially, had applied to become a stewardess in 1937. This last throw of the dice was a desperate attempt to avoid being forced by a well-meaning but domineering friend to accept a grim job as the resident matron of a hostel for factory girls in Hoxton. Her 'maiden voyage' as a rookie stewardess, sailing from London to Argentina, had engendered in her a love for her new career: 'Someone had told me that after two trips, sea-life exerts so powerful a magnetism that one can never leave it and be content. Well, one had been strong enough for me. That drop of sea water was fizzing in my veins. I was going back to the sea.'[4] Maida wanted to work as a stewardess despite the hostilities and restrictions of wartime. She was disappointed at not being recruited to join the crew of a British-owned Blue Star Line passenger vessel bearing interned Italians and Germans, who were being taken to Canada. Within weeks she learned that the ship, the *Arandora Star*, had been torpedoed by a U-boat in the Atlantic on 2 July 1940; 868 people survived but 865 had died. However, Maida managed to get a job in 1940, escorting women and children evacuees, sailing from London to New Zealand, to escape the war.

Maida's passenger ship was to sail across the Atlantic, through the Panama Canal, then across the Pacific. Civilian ships sailed in convoys with a naval escort, zigzagging across the oceans in order to deter submarine attacks. The accompanying naval vessels were equipped with depth-charges to destroy enemy submarines once detected, but often the first indication of one in the vicinity was a torpedo strike on one of the convoy. So dangerous and omnipresent was the threat from U-boats that mothers were told never to leave their children below decks in case of a sudden torpedo attack. Passengers had to carry their life jackets wherever they went on board, and staff had to wear them at all times. For stewardesses it was difficult to make beds and perform their other duties as life jackets were unwieldly garments reaching from neck to hips, restricting bending. Boat drills were frequent; it was vital that all on board knew how to escape from the ship in lifeboats in case of attack. By the time they reached the Panama Canal, the radio news carried nightly stories about the London Blitz.

One of Maida's fellow stewardesses on this voyage was called Nancy Bell, and they became good friends. There were many children, both accompanied and unaccompanied, on the ship and constant vigilance was essential; there were so many dangers for mobile, fearless youngsters. One juvenile, Jimmy, travelling with his absent-minded mother, was spotted one afternoon on the main deck, standing on the ship's rail, balancing on the balls of his feet as the ship rolled and pitched across the ocean. Onlookers froze in horror; Nancy Bell had the presence of mind to creep forward towards him. She grabbed Jimmy, catching him just as he let go of the funnel-stay, pulling him to the deck. Jimmy protested vociferously; Nancy handed him over to his chastened mother, then went to Maida's cabin to have hysterics and a tot of Scotch. As Maida observed, 'She was one of those people who rise to the occasion and collapse at the proper time, after the event.'[5]

On their return journey the Pacific was appropriately peaceful, but once through the Panama Canal they turned towards the war zone of the North Atlantic. From Nova Scotia they set out as part of a convoy, escorted by the armed cruiser *Jervis Bay*. On the afternoon of 5 November 1940, Maida and Nancy were standing at the rail looking at the convoy, when Nancy had a premonition that they were about to face a terrifying ordeal. Her occasional claims to second sight were given credence by many of the crew, who tended to be superstitious, and she had startled and impressed Maida by predicting 'a great slaughter from the skies' a day or two before the Blitz started. The conversation turned to disaster preparations; a fellow stewardess, Barbara, announced she planned to change into trousers if the ship was torpedoed, as skirts would not be 'decent' if they had to climb down ropes or jump into the sea. Nancy pointed out that jumping into water often wrenched off trousers, and that if Barbara feared exposure, she would have to stay bobbing about in the water, rather than risk the shame of being rescued half-naked. Resourceful Nancy planned to be wearing her bathing suit and her furs if she abandoned ship.

At 5 p.m. the same day, the convoy was attacked by a German raider, the *Admiral Scheer*, and the *Jervis Bay* attempted to protect the Allied vessels by engaging with the enemy in order to give the convoy time to scatter. Maida, Nancy and the passengers watched in horror from the deck.

Outgunned, the *Jervis Bay* was quickly consumed by fire and sank, still firing salvoes as it went down. The *Admiral Scheer* circled the remaining unprotected ships, and attacked four of them, sinking three. Then it approached Maida and Nancy's ship. Two mighty crashes rocked their vessel. A sheet of flame rose from the stern, and they assumed they had been hit. In fact, the chief gunner had jettisoned the anti-submarine depths charges in the bows to minimise the risk of damage, and one had exploded spectacularly but harmlessly underwater. The Germans assumed they had fatally hit the ship, and went in search of other prey, allowing Maida's captain valuable time to take evasive action.

The crew and gunners remained at battle stations, while passengers and stewardesses sheltered in the dim and packed alleyway just below the main deck, wearing their life jackets and clutching their 'shipwreck' bags, awaiting instructions to take to the lifeboats. The stewardesses removed their distinctive white headscarves to make them less visible as targets, because the Germans had been known to machine-gun survivors in boats or in the water. A tense wait ensued, as the ship dodged and weaved, engines straining, trying to stay out of the range of the *Admiral Scheer*, which continued to send shells thundering in their direction as darkness fell. A moment's light relief was provided by the youngest stewardess, 'Jock' McCrae, describing their assailant as a 'packet bottleship', a spoonerism that provoked brittle laughter. To keep their spirits up, they sang rousing choruses of 'Roll Out the Barrel', competing with the gradually receding sounds of shells. Almost incredibly, it appeared they had managed to escape. After a tense and largely sleepless night, dawn broke and they were alone on the ocean, as the rest of the convoy had followed orders and scattered. As a single unescorted ship, they were now in dire peril from German raiders, and there was the ever-present menace of submarines. No wireless messages could be transmitted in case they were picked up by the enemy, giving away the ship's position. Later, Maida learned that the Germans had officially announced the sinking of her ship, as part of the havoc they had wreaked on the convoy.

For a fortnight, the ship travelled alone across the Atlantic, 'as solitary as the Ark' in Maida's words, heading for Britain. Those on board kept busy with a regime of frenzied cleaning and polishing. By a miracle, they were not detected by the enemy; the ship crept into Liverpool Bay just as a severe night-time blitz started, so they headed for Milford Haven, only to find it also under aerial bombardment from the Luftwaffe. Inching into Barry Docks, they saw that the harbour was littered with floating mines, so they headed across the Bristol Channel and finally found sanctuary in Avonmouth. Passengers and cargo were unloaded safely; they had had an extraordinary journey, from New Zealand to Somerset, and a miraculous escape.

By the summer of 1940 German forces had swept through Europe and France had fallen. In the skies, the Battle of Britain raged as Allied planes desperately fought the German Luftwaffe for air supremacy. U-boat bases were established at Atlantic ports much nearer to the transatlantic shipping routes, and submarines went hunting Allied shipping, to devastating effect.

The possible invasion of Britain by German forces seemed an imminent threat, and many families feared for their children, especially as nightly aerial bombing increased over the major cities. Some who could afford it, such as Lady Diana Cooper and Vera Brittain, sent their children abroad to friends or relatives, but for most this was not an option. In June 1940 the British government founded the Children's Overseas Reception Board (CORB), evacuating children to the relative safety of the British dominions, including Canada, Australia, New Zealand and South Africa. It was a popular scheme: within a week of its launch more than 210,000 applications were received for just 20,000 places. Disadvantaged families from the areas deemed most at risk were prioritised. Children, known as called 'seavacs', would be looked after during voyages by volunteer adult escorts, and families clamoured to get their children sent to safety, though with mixed feelings. One teacher commented sadly, 'We are sending away our crown jewels', and it was common to see files of small children trudging along railway platforms, clutching their few personal belongings, gas masks slung over their shoulders.

In June 1940, CORB employed Edith Sowerbutts. An able and competent administrator now in her late forties, with a genuine commitment to welfare, Edith was perfect for the job. She knew the North Atlantic merchant fleet and its ships, she had worked in emigration offices, and she was familiar with Canada, America and Australia. Her role was to find suitable men and women to act as escorts to the thousands of children to be despatched overseas. London was badly bombed day and night at this time, but the CORB team, housed in Thomas Cook's offices near Piccadilly, worked on processing applications from thousands of volunteers, surrounded by teetering stacks of files and paperwork.

Crossing the Atlantic was not without risks, even for juvenile civilians. On 29 August 1940 the *Volendam* left Liverpool carrying 320 children under the CORB scheme. On its second day out it was torpedoed seventy miles off the Donegal coast of Ireland. They were lucky; the ship was still close to land, it was a calm night and the ship's lifeboats were safely deployed, so all the children survived.

However, another ship, the *City of Benares* was not so fortunate. Though usually based in London, in September 1940 Edith travelled to Liverpool to oversee the embarkation of ninety CORB children on a passenger ship bound for Canada. Edith arrived in Liverpool and checked in to the Adelphi Hotel in time to avoid an air raid. She always travelled light – just a suit, coat, spare underwear, nightdress and umbrella. Her family had no idea of her whereabouts or her job as she had signed the Official Secrets Act. Edith did not even know the name of the ship going to Canada, but she had an identifying code number for it. She carried typed details of the ninety children due to embark, hidden underneath some knitting in a stout paper carrier bag from a department store. Her homely disguise belied her official and highly confidential mission, though she also thought it unlikely she would ever be mistaken for a beautiful spy.

Edith met the escorts and their excited young charges at Fazakerley Cottage Homes to go through the formalities and have final medical checks. All the children had been provided with warm woollen winter clothes for the voyage by Marks and Spencer, as part of their support for the war effort. One five-year-old, Leonard Grimmond, was so excited that he was sick down Edith's coat, but she forgave him – he was the youngest of five siblings, and their family had been bombed out in London. The children were to join the *City of Benares*, a ship that was now under the control of Cunard White Star Line. Edith also met an old friend, Mrs Whatmore, former senior bath attendant on the *Queen Mary*, who was in charge of linen for the voyage. The two former colleagues had lunch on board the ship with the evacuees before it departed. Edith described the joyous mood:

> It was like a school outing, only lots better – there was a carnival atmosphere. The children were so thrilled, and delighted with the varied menu in the big dining saloon. We stayed on board all afternoon, and were able to watch the lifeboat drill. The children, wearing lifebelts, were thoroughly instructed in all procedures and assisted in and out of the lifeboats by Lascar seamen and Goanese stewards. The scene is etched in my memory.[6]

The *City of Benares* left Liverpool on 13 September under the command of Captain Nicholl, as part of an escorted convoy of nineteen ships bound for Canada. Six hundred miles out to sea, the escort ships departed to meet another convoy heading for Europe. The *City of Benares* was sailing in darkness when it was torpedoed by a German U-boat on 17 September 1940. It was 10 p.m., heavy seas were running, and it was bitterly cold. The ship went down in twenty minutes. Many children were immediately drowned; others found themselves adrift in the dark in partially swamped lifeboats manned by distraught, sodden and scared strangers. Those who had managed to get into boats mostly succumbed overnight to exposure, as they were wearing nothing more than thin nightclothes.

Edith remained unaware of the tragedy for days. She was in Glasgow with another group of evacuees when she was unexpectedly visited by her boss, Geoffrey Shakespeare. He broke the news that the *City of Benares* had been sunk, but Edith was sworn to secrecy until each and every bereaved family had been visited and told of its loss. Out of ninety CORB children, only three girls and three boys had then been rescued. Edith and Mr Shakespeare were taken out to a destroyer at Gourock to meet one of the adult survivors, and the three boys; the three girls had already been taken to hospital with an escort, Mrs Lilian Rose Towns, a schoolteacher. Edith was haunted by the sight of the shocked and dazed faces of the survivors, walking as though in a trance.

Edith was appalled to think that only six of 'her' CORB children had survived, but eight days after the sinking, another lifeboat was sighted drifting on the ocean by a Sunderland flying boat. Piano teacher Mary Cornish and Father Roderic Sullivan, a Catholic priest, were rescued, with six more boys. Mary Cornish was awarded the OBE for her outstanding heroism. It was some small solace for Edith to discover that two siblings who she had particularly liked, Bessie and Louis Walder, fifteen and ten years old, had both survived. Louis was in one boat, while his sister managed to cling to an upturned lifeboat until rescued. Altogether 294 lives were lost, including 81 children. Among them were Michael Brooker and Patricia Allan, who had survived the sinking of the *Volendam* weeks before. While Edith gathered together the juvenile survivors, worried parents who had expected to hear that the ship had arrived safely in Canada besieged the organisation. Nothing could be said publicly until each bereaved family had been personally visited and informed of their losses. Edith particularly remembered a little girl from Liverpool, Aileen Murphy, whose father had put her on the ship himself – he was in an RNVR uniform. He phoned Edith seeking news, but she was sworn to secrecy, even though she knew Aileen was not among the few survivors.

Edith was now faced with the harrowing business of reuniting traumatised survivors with their families. Two youngsters, Rex Thorne, aged thirteen, and Jack Keeley, eight, had each lost a sister in the disaster. Edith accompanied them on the sleeper train back to London. She recalled: 'they talked together like two little old men, asking each other, '"When did you last see your sister?"'[7]

When the news of the catastrophe broke, so great was the public horror that the CORB evacuation initiative was abandoned. Questions were asked about why the protective escort did not stay with the Children's Ship, when U-boats were known to be operating beyond the west coast of Ireland. Families decided they would rather risk the bombs together than send their children across the oceans alone to face such hazards. All 600 children on board ships about to sail were returned to their homes. CORB was largely wound up; Edith left the organisation in November 1940 and became involved in welfare work in London. Based in Hampstead in north London, she was put in charge of requisitioning empty buildings to rehouse people who had been bombed out elsewhere.

Towards the end of the Second World War, Edith was invited to Liverpool by Miss Prescott, the lady superintendent for Cunard. Miss Prescott wanted to discuss her ambitions for re-staffing Cunard passenger ships with female staff when peace returned. Many experienced stewardesses had been hurriedly laid off when the war started, and inevitably they had found new careers ashore, which they were now reluctant to leave in order to resume a life afloat. Edith was one of them; after welfare work she had moved to personnel work, and was now living with her ailing sister. Miss Prescott had hoped she could persuade Edith to return to Cunard as a senior stewardess on a prestigious vessel, but Edith declined. She never forgot her involvement in the *City of Benares* tragedy, and did not resume her career at sea, even after peace was restored.

The story of the sinking of the *City of Benares* reverberated round the world and caused international outrage. The tragedy had a profound effect on Austrian-born Hedy Lamarr, who had escaped an unhappy marriage and the growing political menace of National Socialism in 1937 by crossing the Atlantic on the *Normandie*. In the process she reinvented herself as a glamorous Hollywood film star. Hedy's family were Jewish, and by autumn 1940 her widowed mother, Trude Kiesler, who had also left Austria to escape the Nazis, was becalmed in London, hoping to get a visa for America and passage on a liner to join her daughter. The sinking of the *City of Benares*, and the possibility that other passenger ships could meet the same fate, galvanised the inventor in Hedy. She collaborated with a friend and neighbour, the Hollywood-based composer George Antheil. They both had personal incentives to assist the Allied war effort: Hedy wanted to ensure her mother's safe passage to the States, while George was mourning his brother Henry, a diplomatic courier, who had died on 14 June 1940, when his plane was shot down over the Baltic Sea.

Their joint invention was a blend of Hedy's inspiration and George's development. They wanted to invent an improved radio-controlled torpedo that could be used by Allied forces to attack German submarines. Hedy and George invented a secret method of avoiding jamming, which might interfere with their torpedo and send it off course. By manipulating radio frequencies at irregular intervals between transmission and reception, their system made a secure communications channel which it was impossible for outside agencies to disrupt. George and Hedy patented the invention in 1941 and provided the details to the US Navy, but at the time it was shelved as having no practical application.

Meanwhile Hedy joined the war effort, touring the States and raising $25 million in sales of War Bonds (worth some $343 million today). Her mother reached the USA safely, and eventually joined her in California

In truth, Hedy and George's invention was far in advance of its time, but during the Cuban Missile Crisis of the early 1960s their system was used on US Navy ships. It subsequently had many military applications, and what became known as 'spread spectrum' technology later revolutionised the digital communications explosion, as it was the basis of fax machines, cellphone technology and other wireless operations. It was not until 1997 that Lamarr and Antheil's extraordinary invention received official recognition, when they jointly won the Electronic Frontier Foundation Pioneer Award. In the same year, Hedy was the first female recipient of the BULBIE Gnass Spirit of Achievement Award, given to lifelong inventors, and known as the the equivalent of the Oscars for inventors.

Crossing the Atlantic on a blacked-out Allied ship during the Second World War was a fraught and nerve-racking experience because of the ever-present threat of submarines and gun attacks from German ships. For civilians who had inadvertently found themselves on the wrong continent it was also extremely difficult to obtain a ticket, and any voyage involved circuitous and lengthy diversions, which added to the heightened sense of combined tension and frustration recorded by a number of female travellers.

After spending nearly twenty years living abroad, writer and activist Nancy Cunard, the great-granddaughter of the shipping line's founder, was travelling round South America when France fell to Germany in 1940. Restless and troubled, and estranged from her family, she decided to return to embattled London, where she had spent her teenage years. She undertook a succession of journeys, 'hopping' from Chile via Mexico to the West Indies, on to Cuba, north to New York and finally got a berth on a ship going to Glasgow. Her arduous travels took almost a year, and she didn't arrive back in London until 23 August 1941.

As usual, Nancy wrote obsessively while travelling, recounting the dreadful conditions in which she had seen African-American workers living, the Civil War in Spain and the Spanish refugees confined in French concentration camps. Human rights were her great passion. On arrival in New York from Cuba on the *Marqués de Comillas*, she could not go ashore because she did not have an American in-transit visa, but she was allowed on to Ellis Island at her own request, despite the justice department forbidding her from landing on American soil. There she spent five days mustering her influential American friends and chivvying immigration officials into providing asylum for a Chinese author who had been facing deportation back to Chiang Kai-shek's China, where he faced certain death. Her appeal was successful, and he was allowed to stay in the States. The authorities were not sorry to see her go; Nancy's confrontational championing of black rights in the 1930s, from her flagrant inter-racial relationship with African-American pianist Henry Crowder to her fund-raising in defence of the Scottsboro Boys – nine African-American teenage boys who had been falsely accused of raping two white women in Alabama in 1931 – had left her *persona non grata*, despite her illustrious surname.

The British ship sailing from New York to Glasgow via Newfoundland on which Nancy finally obtained a passage was part of a convoy of vessels taking passengers, food and supplies to Britain. She was unaware their flotilla was to be joined by the HMS *Prince of Wales*, carrying Prime Minister

Winston Churchill. He had recently met President Roosevelt off the coast of Newfoundland, and his presence had to be kept secret from the Germans at all costs. It was difficult to ascertain how many ships were in the convoy but numbers varied between forty-four and seventy-two.

Nancy kept a vivid journal of the transatlantic voyage, which lasted from 31 July to 21 August 1941. Wartime restrictions were imposed; no smoking was allowed on deck, which Nancy found irksome, as she smoked constantly. The portholes were screwed shut and there were blackout shutters covering every window, in case of submarines, battleships or planes. The passengers learned their lifeboat drills, and the radio news featured carefully edited tales of the Blitz, and accounts of British heroism. On 15 August she saw on the starboard side:

> a monumental battleship with a full and stately flurry of spray at her prow, her guns pointing skyward, a great bridge between her two funnels. It is Churchill, returning from his talk with Roosevelt on the *Prince of Wales*, six or seven cruisers accompanying them. This majesty passes us quickly, crosses in/out and comes by the right and the ship's other side – to see and be seen. Churchill's showmanship – a delight to us all.[8]

The following night there were air raids and surface attacks, followed by anti-aircraft fire. The convoy reached Iceland, and the onward journey to Scotland was hazardous with icebergs and mines, which were detonated by an escorting vessel. The seas were rough and there was a great deal of seasickness and illness on board. On arrival in Glasgow, she wrote: 'How modest and even almost unimpressed are those who have charge of this great procession over the ocean ...

Yes, remarkable … the *Might* of Britain. Inspiring.'⁹ Nancy returned to London by train and stayed there for the next three years, dodging the bombs and living in straitened circumstances in bedsits, just a few minutes' walk from her estranged mother, who held court in a grandly furnished but cramped suite on the seventh floor of the Dorchester Hotel.

There were many tales of individual bravery by women on the Atlantic during the Second World War. Victoria Drummond, the first woman to qualify as a marine engineer in the 1920s, had given up her job ashore and joined the war effort, becoming a second engineer in the Women's Mercantile Naval Reserve, and assisted with the evacuation at Dunkirk in 1940. According to the *Evening Standard*, Councillor William Lockyer, the Mayor of Lambeth, who was a friend of the family, said: 'She has been bombed during her voyages, but all she says about it is "They are such bad shots." Once a submarine attacked her ship with torpedoes, but missed.'¹⁰

Victoria Drummond was the second engineer and the only woman aboard the cargo ship *Bonita* when the ship was attacked mid-Atlantic by a German plane. They were 400 miles from land. The first salvo from the bomber threw Victoria against the levers on the control platform in the engine room, and nearly stunned her. When the stokehold and engine room staff had done all they could to get an extra knot or two out of the ship, she ordered them above, giving them a chance of survival while she stayed alone below, where she had little hope of escape. Each time the plane bore down on the ship to drop its bombs and strafe the decks with bullets, the captain would take evasive action, a finely judged manoeuvre requiring a surge of extra power from the engine room. With one hand holding down the throttle control, while the enemy plane roared overhead, Victoria coaxed the engines from nine to an unprecedented twelve and a half knots, just enough to avoid a direct hit. The attack continued for thirty-five minutes, and eventually the frustrated enemy pilot, short on fuel, was forced to retire. One of Victoria's fellow officer colleagues left a vivid account of the conditions in the engine room during the bombardment:

> It must have been hell down there. Two cast iron pipes were fractured, electric wires parted, tubes broken, and joints started, but her iron body and mighty heart stood it. The main injection pipe just above her head had started a joint and scalding steam whizzed past her head. With anyone less skilled down there that pipe would have burst under the extra pressure, but she nursed it through the explosion of each salvo, easing down when she judged from the nearness of the plane's engines that the bombs were about to fall, holding on for all she was worth to a stanchion as they burst and then opening up the steam again. If the pipe had gone, we would have stopped and it would have been all up. By getting the speed it gave the helm a chance to move the clumsy hulk, and literally every second mattered in the swing.
>
> I saw her once during the action when I had to dodge along to the W.T. room and looked down the skylight, hoping to be able to shout a few words of cheer to her. She was standing on the control platform, one long arm stretched above her head and her hand holding down the spoke of the throttle control as if trying by her touch to urge another pound of steam through the straining pipes. Her face, as expressionless as the bulkhead behind her, and as ghastly white in colour, was turned up towards the sunlight, but she didn't see me. From the top of her

forehead down her face, completely closing one eye, trickled a wide black streak of fuel oil from a strained joint. That alone must have been agony. She had jammed her ears at first with oily waste to deaden the concussion and then tore it out again, for fear she would not hear some vital order from the bridge – not knowing that all connection with the bridge was cut. She was about all in at the end, but within an hour was full of beans and larking about and picking up spent bullets and splinters. All round her, by the way, the platform was littered with bullets that came down from the skylight.[11]

Victoria Drummond was commended for her bravery, which on 9 July 1941 the *London Gazette* recorded 'was an inspiration in the ship's company, and her devotion to duty prevented more serious damage to the vessel'. Edith Sowerbutts had long followed Victoria's pioneering career, and later wrote in her characteristic forthright manner: 'Her record is fantastic – forty years at sea; she was made an MBE in the Second World War. Yet, the young engineers who served in my ships opined that she was "crackers" – had she joined us, she would have been in for a rough time.'[12]

Victoria Drummond was a remarkable seafaring woman doing a job otherwise exclusively occupied by men, but there were many other wartime acts of bravery among women who found themselves working at sea on more equal terms. Medical Officer Dr Adeline Nancy Miller was on the armed troopship *Britannia*, owned by Anchor Line of Glasgow, when it was shelled by a German raider, the *Thor*, on 25 March 1941 just off the west coast of Africa. As the *Britannia* was losing speed, the captain gave orders to abandon ship, and a message to this effect was conveyed to their attacker. But the

Thor continued to bombard the *Britannia*, holing many of the lifeboats. As the *Britannia* began to sink, the German raider sailed away. Dr Miller calmly attended to the wounded and the dying. She managed to save many lives, and the survivors took to the remaining lifeboats. One single lifeboat certified to hold fifty-eight people contained eighty-four, and it took twenty-two days before they made landfall. The lifeboat in which Dr Miller spent five and a half days, caring for the sick and wounded, was picked up by a Spanish cargo steamer, the *Bachi*. By a coincidence the *Bachi* was then intercepted by the *Cicilia*, the ship on which Dr Miller's father Thomas was employed as ship's surgeon. He had heard the radio reports of the sinking of the *Britannia* nearly a week before and had believed his daughter was dead. His relief when Adeline climbed on board and embraced him was profound. She was awarded the Lloyd's Medal and the MBE.

The steely determination of the merchant fleet during six long years of war made Fortress Britain a reality, and while passenger travel across the Atlantic was vastly reduced, some civilians did still make the crossing, in convoys, sometimes using ships belonging to neutral countries. A small number of British seafaring women also worked on the ships in various capacities on a voluntary basis, and their hazardous role was commended by influential journalist Hannen Swaffer. He cited an indignant forty-seven-year-old stewardess called Margaret Thomas, from Edmonton:

> Does nobody realise that stewardesses and ships' nurses are still regularly putting to sea, sharing the ever-present dangers equally with the men, and without the physique of the men with which to meet those dangers? Yet so many stewardesses have volunteered for active service

that we have to wait months for a ship. And we know what we are up against, when we have been once. But no mention is ever made of the women in the Merchant Service, still carrying on in the face of appalling danger, and under intense nervous strain. It is high time that somebody rectified that omission, and that everyone real-ised that British women are still sailing the seas under the 'Red Duster'.[13]

The majority of stewardesses were laid off, and many of them – such as Dorothy Scobie – joined the WRNS, the Women's Royal Naval Service, along with women who had no previous maritime experience. The WRNS had been disbanded after the end of the Great War, but was now reac-tivated by the Admiralty. Its motto had been 'Never at Sea', a phrase open to two interpretations: while it implied they were resourceful and able, it was also accurate because initially they were only deployed on land, in support roles. A BBC Radio broadcast by the Duchess of Kent in January 1941 encouraged women to join the WRNS as cooks, wait-resses, clerks, bookkeepers and typists. There were also more senior roles in ciphering, signalling and wireless telegraphy, and it was stressed that every job taken by a woman enabled a male naval officer to be released for more active duties with the fleet. The appeal was successful and women joined the WRNS for training of all kinds, eventually taking on more than 200 roles. They were not allowed to work on ships in active combat zones, though a number of them did operate small harbour launches and tugs, or acted as pilots for large vessels. Their contribution to the war effort was graphically brought to public notice when twenty-two WRNS personnel and one nursing sister were drowned in August 1941. The SS

Aguila was taking them from Liverpool to Gibraltar, where they had volunteered to serve as cipher officers and wireless operators. The *Aguila* was part of a convoy of twenty-one vessels, and it was attacked by a 'wolfpack' of twelve German U-boats, each of which took turns to pick off a total of fourteen ships. The *Aguila* took a direct hit from a torpedo and sank within ninety seconds. When the news broke, other WRNS staff donated a day's pay to a memorial fund, and the £4,000 raised funded an escort boat, the HMS *Wren*, to accompany future convoys.

The WRNS worked ashore or close to port until 1943, when thirty women were deployed on former passenger ships, taking high-level figures to secret meetings and international conferences. The WRNS personnel on board encrypted and decrypted top-secret messages, but were always subordinate to their male colleagues. However, being a Wren did provide some women with the opportunities to train in what had traditionally been men's roles, and after D-Day in 1944 they followed the Allied forces into liberated Europe, providing support services and logistics. One of them was a field tele-printer operator, who was landed with her fellow Wrens in Normandy, and their unit set up in newly-liberated Paris. She became Laura Ashley, the world-famous textile and fashion designer. By the end of the war, more than 75,000 women had served in some capacity in the WRNS, and 303 of them had died on active service. In recognition, the navy retained a regular force of about 3,000 women in peacetime, though they did not serve regularly at sea until the 1970s.

As in the First World War, British women were keen to help the war effort, and they were often employed in ship-yards. In 1943 *The Times* approvingly reported that women were fulfilling 114 different jobs in shipyards, evidence that female workers could be employed on any task that did not require either years of specialist training or considerable physical strength:

> As electricians, painters, tool maintenance hands, sheet metal workers, wiremen and on the many other processes which go to make ships, women are doing excellent work, and their skill at welding has been generally acknowledged ... At a time when few men are available, it is interesting to read that all-women gangs, under a woman supervisor produce the best results. In mixed gangs the men are, it is stated, inclined to use the women as assistants or labourers, whereas when women are working alone their self confidence and enthusiasm grow, and they become more effective.[14]

By summer 1944 there were some 13,000 women working in marine engineering in British shipyards, and they were the counterparts of America's famous Rosie the Riveter, helping to build and repair ships. The entry of America into the war following the bombing of Pearl Harbor was decisive in changing the theatre of war in Europe and on the Atlantic. Previously the *Queen Mary* and *Queen Elizabeth* had been bringing Australian troops to fight; but in 1942 they were back on the transatlantic run, bringing men, munitions and machines to Europe, at full speed.

As the fortunes of the war gradually changed, the Allies began to prepare for the invasion to be known as D-Day. Up to 15,000 service personnel per voyage would be packed aboard the *Queen Mary* or *Queen Elizabeth*, occupying every available space. So congested were the corridors and public spaces that a one-way system was introduced for passengers. Commodore Sir James Bisset noted that the *Mary* was so difficult to handle under such circumstances that he was concerned for its stability. On one voyage the ship carried 16,500 people, which is still a record today. All told, the *Queen Mary* and *Queen Elizabeth* each made nearly thirty trips eastbound, carrying American soldiers to Europe, and prisoners of war westwards to internment.

The catering staff were constantly cooking to provide two substantial meals a day for all on board, and dining was in shifts. The whole trip still usually took five days and the troops passed their time playing card games, dice and poker, even though, in theory, gambling was forbidden. WAACs, the American Women's Army Auxiliary Corps, were also transported on the *Queen Mary*. They were strictly segregated from the male passengers, and everyone was barred from smoking on deck, in order to maintain the strict blackout.

On three occasions the *Queen Mary* carried the British Prime Minister, Winston Churchill, in conditions of utmost secrecy, to meet President Roosevelt and discuss the progress of the war, the arrangements for D-day and its aftermath. Naked flames were not allowed in cabins at any time, but special allowance was made for the British premier to have a candle burning constantly so he could smoke his trademark cigars. He was listed on the passenger manifest as Colonel Warden to conceal his identity.

In the summer of 1944, as the preparations were finalised for the Allies to invade Europe, there was one American woman who was on the wrong side of the Atlantic, and she was willing to resort to desperate measures to cross the ocean and witness the battle. Martha Gellhorn, inveterate transatlantic traveller and war journalist, had first managed to get to Europe by writing the copy for a brochure for the Holland-American Line, a Dutch passenger shipping firm. She became a foreign correspondent, travelling through Nazi Germany, Spain during the Civil War, and experiencing the Blitz in London. She wrote for a number of illustrated American magazines, especially *Collier's*. Martha Gellhorn had married the writer Ernest Hemingway and they were living in Cuba. She wanted to cover the invasion of France by the Allied forces in 1944 and so needed to get to Britain to join the invasion fleet, but her travel options were limited.

Resourceful Martha managed to secure a passage on one of the most dangerous vessels afloat, a cargo ship heading for Liverpool, packed with dynamite. There were forty-five Norwegian sailors on board; the captain and first mate had a rudimentary grasp of English, but otherwise she was unable to communicate for eighteen long days. The deck was covered with small, amphibious personnel carriers, so there was almost no space in which the solitary passenger could stretch her legs. The hold was filled with high explosives, and there were no lifeboats; there was no point with such a hazardous cargo, as the slightest accident would engulf the vessel in a fireball. Smoking was forbidden, though the captain permitted Martha to smoke in her cabin if she used a bowl full of water as an ashtray. The food was appalling, it was extremely cold, and there was no alcohol on board, for safety reasons. While at sea, Martha's ship avoided icebergs, dodged submarines and had gunnery practice. Fog descended, and the captain was concerned about the unpredictable manoeuvres of the interweaving Liberty ships accompanying the convoy, muttering that 'they try to handle them like a taxi'.

On the grounds of her gender, the British government would not allow Martha Gellhorn to join the 558 writers, radio journalists and photographers provided with official press credentials so that they could follow the Allied troops storming the Normandy beaches on 6 June 1944. No women were to be part of the press corps covering this dangerous military venture. To her intense irritation, Martha's estranged husband, Ernest Hemingway, was the single representative commissioned by her previous employers, *Collier's*, to cover the D-Day landings. Undaunted, Martha managed to get aboard a hospital ship, and stowed away in a bathroom. She found a nurse's uniform and changed into it, and in disguise she proved rather useful to the crew; she was the daughter of a surgeon, and could speak fluent German and French, so could act as an interpreter as well as carry out rudimentary nursing duties. In conditions of great secrecy, the ship – one of nine medical vessels to sail with the D-Day fleet of 100,000 troops and 30,000 vehicles – joined the invasion force heading across the English Channel to the French coast. The crew of Martha's ship were English, but the medical staff were American, comprising four doctors, fourteen orderlies and six nurses: 'from Texas and Michigan and California and Wisconsin, and three weeks ago they were in the USA completing their training for the overseas assignment. They had been prepared to work on a hospital train ... instead of which they found themselves on a ship, and they were about to move across the dark, cold green water of the Channel.'[15]

When they reached the coast of Normandy, Martha was astounded that there could be so many diverse vessels in one place. Fierce battles ensued as the Allied troops forced their way ashore to face German bombardments, and before long water ambulances were bringing wounded combatants of all nationalities out to the hospital ships moored off the coast. The medical staff were working at full stretch – the patients had to be lifted aboard, triaged, sent for surgery or treated for their wounds. Their clothes and boots had to be cut off them, most of them hadn't eaten for two days, and they were desperate for water, food, coffee, cigarettes and pain relief. Martha spoke to many of them, in German and French, realising they were extremely young and very frightened; she carefully explained to one young German that the orderlies dare not move him as he might bleed to death. There were more wounded stranded on the beach, and Martha volunteered to go ashore as a stretcher-bearer. It was a hazardous business; her party landed at dusk and trekked up a beach over pebbles the size of melons to a Red Cross tent. Martha helped supervise the transfer of the wounded to craft that could take them to the hospital ship as soon as the tides were right. Once back on the ship, with every bunk filled with injured servicemen of many nationalities, they sailed for the English coast where the wounded were taken to hospitals on land. 'Made it' was the terse but heartfelt verdict of the chief medical officer.

Martha exchanged her nurse's disguise for her own clothes, left the ship and took the first train to London. She was promptly arrested and detained but escaped during the night, making her way to the home of an RAF pilot she knew. He was flying to Italy the following day, so she hitched a ride with him to the next theatre of war.

Her account of the D-Day landings was vivid and personal. She focused on the human tales of bravery and endurance, but what is also remarkable is her own commitment to getting the story by travelling the Atlantic on a dynamite ship then stowing away on board a hospital ship. This was not the first eyewitness account of war at close quarters related by Martha Gellhorn, and it would not be her last, but it was a milestone in her remarkable sixty-year career.

Following D-Day, while the war in Europe intensified,

transatlantic passenger numbers gradually increased. Even though conditions were still dangerous, there were renewed opportunities for some merchant seawomen to resume their former careers. Maida Nixson, whose accidental career as a stewardess had begun in 1937, had longed to go back to sea, and had already risked enemy torpedoes and death by drowning, escorting evacuated mothers and children around the globe to safer countries. The WRNS could only offer her domestic work in port, so Maida became an assistant nurse in a hospital. However, during the last year of the war she found a job as a stewardess again, this time dealing with war brides and their offspring.

In late June 1944 Maida set sail in a largely empty passenger ship from Liverpool to New York, to help organise the transport and resettlement of American and Canadian wives of British airmen. The pilots and ground crew had been sent to the States and Canada for training in the early years of the war, and those who had married local girls wanted to be reunited with them, now that the risk from German raiders and submarines was so diminished. As usual, the ship zigzagged across the Atlantic, so progress was slow; meanwhile radio news reports spoke of the destructive flying bomb attacks on Britain. Maida's five fellow stewardesses on this trip included the widow of a steward who had lost his life, along with many of his shipmates, in their valiant attempts to save children trapped below decks when their vessel was torpedoed.

On arrival in New York to pick up the American brides and offspring, the crew were amazed by the contrast with the darkened and war-ravaged British ports to which they were accustomed: 'The skyline of New York must always be considerably overwhelming, and the sight of that great serrated artificial cliffside a-dazzle with blinding radiance made an almost shocking impact on our nerves. Our eyes accustomed to darkened England refused to credit the glare, we felt inclined to call "Put out those lights!" By night, even more than by day, the city had a look of bizarre unreality.'[16]

On sailing day, the ship was swamped by hundreds of young women, accompanied by small, hot toddlers and crying infants. Someone thrust a large, damp baby into Maida's arms, and disappeared into the crowd. It took Maida some time to locate the mother of this unwanted burden, and when she succeeded, she was virtually accused of kidnap.

Extra bunks had been put into the cabins, and below decks there were dormitories housing twenty-six beds at a time. Conditions were cramped and difficult – six stewardesses each had around 130 passengers to look after and there were 99 infants aboard. Most of the passengers had never travelled before and, though nervous, were uncomplaining. The ship was blacked out from dusk to dawn, and it was extremely hot, being July. Meals were eaten in relays, and all passengers had to be off decks by 10 p.m.

Some British people were aboard, having spent the war years in the States, and they were privately dubbed 'bomb-dodgers' by the stewards. On arrival, the women and children were handed over to the RAF. Customs officers discovered a thriving smuggling ring on board, mostly in bird seed, which sold for £1 a pound in savagely rationed Britain, and in Cuban cigars, perhaps easier to understand. A transatlantic voyage in wartime was a great chance for shipboard crew to trade in forbidden or unavailable commodities, and considerable ingenuity was exerted to hide contraband from the customs officials.

Cunard alone had transported 2.473 million people and

9 million tons of cargo during the Second World War. Winston Churchill would later bestow the greatest compliment on Cunard when he remarked that the contribution of the two *Queen*s and *Aquitania* had shortened the war in Europe by at least a year. Winston Churchill had good reason to be profoundly grateful for the crews' professionalism and dedication. In March 1946 he sailed on the *Queen Elizabeth* to the United States to deliver what would come to be known as his 'Iron Curtain speech' in Fulton, Missouri. During the course of that voyage he agreed, at Cunard's request, to write a foreword for a projected war history of *Queen Mary* and *Queen Elizabeth* that Cunard planned to publish:

> Built for the arts of peace and to link the Old World with the New, the 'Queens' challenged the fury of Hitlerism in the Battle of the Atlantic. At a speed never before realized in war, they carried over a million to defend the liberties of civilization. Often whole divisions at a time were moved by each ship. Vital decisions depended upon their ability continuously to elude the enemy, and without their aid the day of final victory must unquestionably have been postponed. To the men who contributed to the success of our operations in the years of peril, and to those who brought these two great ships into existence, the world owes a debt that it will not be easy to measure.[17]

The war in Europe ended with the German surrender to Allied forces, and was marked by the celebrations of VE Day, 8 May 1945. Worldwide, tens of millions were dead, and many more were homeless or displaced. Many European countries were in ruins, with devastated economies no longer in control of their industrial infrastructure. Help from

America in the form of the Marshall Plan was delivered in naval and merchant vessels crossing the Atlantic.

In the months following the end of the war, the great ocean liners helped to repatriate American troops and airmen. With 15,000 General Infantrymen – GIs – aboard (including 4,000 accommodated in sleeping bags in corridors), the *Queen Mary* received a hero's welcome in New York City. Newsreel footage of the ship's arrival shows the decks teeming with cheering, waving servicemen as they pass the Statue of Liberty. The great ships brought home the prisoners of war, repatriated the dispossessed, and carried the stateless to new lives in the New World.

For seafaring women the experiences of wartime had been salutary and a test of their mettle. There were those who had worked aboard the great ships in the 1920s and 1930s, during the golden age of transatlantic travel, when the appeal of the role lay partly in its proximity to celebrity and glamour, and the opportunity to make an independent living. Many of these women, in choosing a career afloat, had demonstrated that they were resourceful and adventurous, and they were willing to use their sea knowledge and their people skills to support the war effort in tougher times. There had been tragedies: Edith Sowerbutts wrote movingly about the sinking of the *City of Benares*, and she did not return to working at sea after that ship was lost. There were also instances of extreme bravery, such as Maida Nixson's determination to escort evacuated women and children to places of safety, and Victoria Drummond's extraordinary courage in sticking at her post to save her ship and fellow crewmates from enemy attack. Intelligence and stubbornness had also characterised many of the women who had travelled as passengers on the ships during wartime: expatriate Nancy Cunard, who took months to reach war-torn London, and Martha Gellhorn, consummate war writer, who claimed a berth on a dynamite ship to cross the Atlantic, then stowed away on a hospital ship to cover the D-Day landings.

For women of all nationalities who had survived the years of conflict, the post-war world offered new challenges and huge decisions. Many of them had met and married men from overseas, and they now faced moving from one continent to another in order to join their husbands. The merchant navy rose to the challenge by recruiting a new generation of women seafarers, primarily to help reunite far-flung families who had been sundered by war.

IO

Romance, Repatriation and Recovery

Bernard and Dora Owens's long and happy marriage started
as a whirlwind wartime romance. He was an American
soldier and, by September 1942, was attached to General
Eisenhower's London headquarters. Late one December night,
as he hurried into a Tube station through the heavy fabric
blackout curtains, he blundered into an attractive English
nurse in a crisp uniform, who berated him for nearly knocking
her over. Apologies followed, and he escorted her in a cab to
her workplace, Mile End Hospital, more than six miles away.
Dora agreed to a date, and they swiftly fell in love. Thirty-five
days after their first encounter, Bernard proposed marriage,
Dora said yes, and they married on 23 March 1943.

Home was a rented apartment in central London, and
their son Michael was born in January 1944. Bernard served
with the American Army in France and Germany, leaving his
wife and son in London, and after the war was over, the
family made plans to move back to America. Bernard sailed
first, and Dora and two-year-old Michael followed on the
USAT *Saturnia*, landing in New York on 27 April 1946. They
moved to Fort Worth, Texas, and Bernard resumed his military career till retiring to San Antonio in 1970. They had
been married for sixty-five years when Dora died in 2008,
aged eighty-eight. Bernard said that he adored his 'English
nurse', and that neither of them ever regretted their chance
encounter years before.

Dora Owens was one of approximately 70,000 British-born GI brides who left their own country to join their husbands in America during the 1940s. A further 150,000 married women and their tiny children made the journey from mainland Europe. Bernard and Dora's heady romance and swift marriage was typical of the era; before D-Day, hundreds of thousands of troops from overseas had been stationed temporarily in the British Isles while waiting for an overseas posting to the main theatres of war. Young servicemen inevitably came into contact with 'the locals', and proximity, curiosity and mutual attraction often led to romance. American servicemen who married foreign nationals were promised free passage back to the United States for their spouses and any children. Many British women also married other foreign nationals, such as Canadians or Australians.

As soon as hostilities ended in 1945, there was a huge incentive to reunite British and European women and their offspring with their repatriated husbands, as it reduced the number of dependent individuals in each country who were subject to stringent rationing and struggling to find accommodation in the wake of the war. However, the process was slow and there was a shortage of passenger berths for transatlantic travel. The brides themselves clamoured to rejoin their husbands, staging organised protest marches outside the US Embassy in London in October 1945, bearing placards proclaiming 'We Want Our Husbands' and 'We Want Ships'.

In December 1945 the US Congress passed the War Brides' Act, granting special status to the foreign-born wives and dependants of US servicemen. They were now exempt from the normal restrictions of the immigrant quotas, and all efforts were made to reunite the families divided by the Atlantic. Operation Diaper Run was the name given by the American

War Department to the initiative, and some thirty American-
and British-owned ships were hastily adapted to accommodate
mothers and babies. Shipping companies contacted many of
their former female seafarers, and offered them posts accompanying the mothers and babies. There was a shortage of
trained female personnel, as so many had been laid off at the
start of the Second World War, and they had since found jobs
on land. Cunard's interim solution was to extend the working
age for stewardesses beyond sixty if they wanted to continue
going to sea, and many stayed on till they were sixty-five or
older, happy to fill a relatively well-paid job they found
convivial.

The first shipment of brides of Operation Diaper Run
boarded the SS *Argentina*, which left Southampton in January
1946. On board were 452 British women, their 173 children
and 1 'war bridegroom'. On arrival in New York, a band
played 'Here Comes the Bride' as the *Argentina* docked at
Pier 90.

Captain Donald Sorrell, captain of the *Queen Mary*,
recalled that his superintendent told him, 'You've taken the
men back, Sorrell, now see what you can do with their
women.' The *Queen Mary* alone transported some 22,000
women and children within seven months.[1] The *Queen Mary*
was refitted in January 1946 in Southampton, and sailed to
New York on 10 February, transporting 1,700 GI brides and
650 infants to their new lives in America. The ship had been
provided with extra laundries, a nursery and a bigger playroom. The area around the first-class swimming pool was
altered to be used as a drying room for nappies. There were
stewardesses, American Red Cross nurses, children's nurses
and welfare officers aboard. Kay Ruddock of the Red Cross
was a war bride escort officer on the *Queen Mary*; she recalled

that it was very sad watching the young women and children saying goodbye to their friends and families as they boarded the ship in Southampton, knowing they were sailing to an unknown future and might not meet again. 'Lovesick, seasick and homesick' was how one GI bride described her conflicting emotions and sensations.

There were all sorts of hazards facing the ship and its charges: it was winter and the North Atlantic could be very stormy, while there were still dangers from floating mines and wreckage. Inside the ship there was a severe shortage of high chairs and these had to be screwed into the deck to secure them, as the *Queen Mary* was notorious for pitching and rolling in rough seas. After six years of severe food rationing in mainland Britain, some women were unable to control their appetites, and over-fed their children too. One proudly boasted to the captain that her two-year-old had just consumed stewed fruit, porridge, eggs and bacon, while a young mother was on her seventeenth bar of unrationed chocolate that morning. The dire consequences of binge-eating while travelling at thirty miles an hour through heaving winter seas were both predictable and unpleasant.

One anonymous stewardess recalled multiple 'baby run' trips on Cunard liners. On her first voyage the passengers were mostly Scottish young women, many of whom had been courted by the American soldiers billeted at a camp near their rather quiet town in the run-up to D-Day. Each young mother was taken aboard, and settled with three others in a four-berth cabin, sleeping in bunks to which metal cribs had been attached for the infants:

> Oh, they were so happy to see all the food they had, and all their bunks. There was all these babies, and they were

so thrilled with all the food they had. They were terribly sick and so were the babies. When we got to New York we helped the mothers take the babies ashore and the American mothers-in-law, you see, all came down to meet them and I remember one American said, 'what beautiful babies, we'll have to send our American children over there to get some colour in their faces', you know. But those girls cried when they left the ship, they felt they were leaving England ... really leaving England for the last time.[2]

Before sailing, all the mothers had been briefed about what to expect in America; after all, they were travelling to a foreign country, and needed to understand the culture, the language and the customs. In June 1945 the Good Housekeeping Institute had published a pamphlet entitled 'A War Bride's Guide to the USA', offering practical tips on contemporary American society, slang and popular culture. As the introduction warned: 'You have undertaken to become an American – just as millions of other people have done before you. Getting to know your adopted country will be an exciting adventure; the future is before you.' British-born women were advised to be pleasant but reticent at first, and to avoid trying to make jokes until they were on familiar ground.

Operation Daddy was the informal name for the Canadian initiative to reunite GI families; some 45,000 British women, accompanied by 15,000 children, were taken by ship to Halifax in Canada to join their menfolk in the late 1940s. A fair proportion were Scottish-born, a reflection of the large number of Canadian troops stationed north of the border during the war years, and it is estimated that one in thirty present-day Canadians is a direct descendant of a 'war-bride' family. Many of the war brides initially suffered from home-sickness and culture shock on arrival in North America, and some marriages inevitably failed, though most survived.

Of course, married women and their babies were not the only people clamouring to cross the Atlantic in both directions. As soon as peace was restored, the depleted shipping companies tried to resume some regular services. Stewardess Maida Nixson was employed on the *Plantano*, sailing between Liverpool and Halifax. Her passengers heading west on the New Year voyage of 1946 were a mixture of war brides and infants, seasoned travellers and displaced European royalty. So pressured were the merchant shipping lines in the immediate aftermath of the war that passengers would take any available ship going from one continent to another, then continue their lengthy onward journey by plane or train. On this voyage there was a contingent of Gold Coast scholars who were going to the USA to study medicine, and were prepared to make a lengthy journey by ship from Africa to Britain, then cross the Atlantic to Canada, before heading overland to their ultimate destinations in American colleges. Next to them in the two suites and adjoining cabins were five members of the Greek royal family, with a nurse and valet, all of whom were heading for Florida, but forced to take a lengthy detour on a ship to Halifax in Nova Scotia. On embarkation, the exiled King of Greece was nearly swept away by a human tsunami of brides and babies surging up the alleyway and sustained a debilitating blow on the royal backside from a passing suitcase. The voyage was notably rough, and the ship arrived in port just in time to see the majestic *Queen Elizabeth*, pride of the Cunard fleet, sweeping out.

Having left bitterly cold and heavily rationed Britain

behind, Maida soon found that the *QE*'s crew had compre-
hensively stripped the Halifax shops of anything desirable,
and the temperature plummeted so quickly that she was
forced to undertake her on-board cleaning duties while
wearing her two fox furs draped over her uniform.[3] There
was some consolation as the *Plantano* docked in New York,
one of Maida's favourite cities, and a complete contrast to
the ports of war-torn Europe. 'Those incredible buildings
stood sharply out against the sparkling cobalt skies of Spring,
and the air was crisp and fresh. And this time, I resolved, I
would see and do all those interesting things I had left undone
before, Central Park, Radio City, a conducted sight-seeing
tour.'[4]

American-born Lady Astor was desperate to get back to
the States to see her family after six long years of war, but
it was impossible to get a passage, even with her wealth and
connections. The only tickets available at short notice were
as passengers on a rather battered Fyffe's banana boat, the
Eros. It was a modest though sturdy cargo vessel, and a far
cry from the *Queen Mary*, the *Aquitania* and Nancy's other
inter-war modes of transport during the golden age of travel.
Impulsive and determined as ever, Nancy booked four tickets,
for herself and her husband Waldorf, his valet (Arthur
Bushell, whose impersonation of Queen Mary was famed
throughout servants' halls all over Britain) and Nancy's stoical
maid, Rose. It took fourteen days to reach New York from
Tilbury instead of the usual week because of the terrible
weather. The ship's cook was a bucolic character who had
been unenthusiastic at seeing the Temperance campaigner's
name on the passenger list, as Nancy was the MP who had
attempted to abolish rum rations for all merchant seamen.
However, so effective was her charm that the entire crew lined

up to sing 'For she's a jolly good fellow' as the Astors disembarked in New York, their first homecoming since the start of the war. Lord and Lady Astor stayed in the luxurious Ritz Carlton, a world away from grey and gritty Britain. Waldorf gave a lunch party for the crew of the *Eros*, which was a great success, and the grumpy chef asked Rose out on a date, which included energetic jiving fuelled by rum and Coca-Cola.

After the end of the hostilities there were millions of people in the wrong place, and they wanted to get home. Others wanted to escape from the countries where they had found themselves when the war ended. Some were prisoners of war, and the great ocean-going ships were put into service to repatriate them. Most of them were men, but there were also women who had been captured during the war and needed repatriation. The author's great-aunt, Margaret Rosina Evans, had been a nurse in Singapore before it fell to the Japanese in February 1942, and for more than three years she was held as a POW in the notorious Changi Jail. Her return voyage to Britain on a passenger ship that sailed via Colombo, Port Said and Gibraltar took more than a month. Margaret finally arrived in Southampton on 17 October 1945. After being interviewed by the immigration authorities, her immediate priorities were to be reunited with her family, have her hair permed and to visit the dentist to have a number of teeth extracted, a legacy of the mistreatment meted out to the camp internees by their captors.

There were also would-be refugees on the move, Europeans who had suffered appallingly as a result of the international conflict, and some of them were now able to take ships to the New World and a new life. Captain Sorrell was put in charge of the Cunard ship *Samaria*, which sailed to pick up 'displaced persons' from Cuxhaven and Bremen, and convey them to Canada. He described how thin and bedraggled the passengers were as they embarked, and how the crew had been instructed to show them compassion and kindness. There were severe communication problems, and not only because of the language difficulties; one elderly woman lay in her bunk day after day and refused all food. The stewardess was mystified; her charge didn't appear to be seasick, and the doctor could find nothing physically wrong with her. It transpired after several days that the woman was desperately hungry, but had been afraid to eat because she had no money to pay for the food. No one had been able to convince her that it was free.

During the voyage Sorrell noted that some of the refugees were starting to recover, and putting on a little weight. Just before the ship docked, two of the passengers asked to see Captain Sorrell. 'I invited them into my cabin and there they stood, proud and shabby, carrying a beautifully carved, colourfully designed wooden box. I was told that they had made it during the voyage. It was inscribed: "To Captain of Ship *Samaria* from Ukrainian emigrants, Oct 21st–Nov 1st 1948". It was one of the most moving tributes I ever received.'[5]

Following the war, *Queen Mary* was refitted for passenger service, and with the *Queen Elizabeth* Cunard resumed the two-ship transatlantic passenger service for which both ships had been built. Cunard had lost nine passenger ships as a result of the war. With the return of peace the company urgently needed to review its vessels and predict coming trends in travel. The *Aquitania*, the last of the pre-Great War liners to survive, and which had made 580 Atlantic crossings, was nearly forty years old, and so it was finally sent to the breakers in 1950. It was the only transatlantic liner to have served in both World Wars, carrying 120,000 troops in the first conflict and 300,000 in the second. Cunard had planned to retire the *Aquitania* in 1940, but wartime necessity saw it pressed into service once again. Having sailed 3 million miles and carried 1.2 million passengers, the ship was remembered fondly on both sides of the Atlantic:

> No ship in modern maritime history has had a more honorable and distinguished record. She was one of the queens of the transatlantic fleet. She was, in two wars, a troopship, a ferry for war brides, and finally a ship of hope for displaced persons. But to those in New York she was something else. She was a recurrent adornment to the waterfront skyline. Her famous four stacks were there after her sister ships disappeared. She was unmistakeable, dignified, proud, often sedate, always the great lady. Those who travelled on her know of her comfort; those who watched her know of her beauty. She was a great and a proud ship.[6]

During the period of post-war austerity, there was considerable demand from Britons wanting to get abroad, especially after the unprecedentedly grim and lengthy winter of 1946–7. Violet Jessop had spent the war years working in a censorship office in Holborn, checking Spanish language post and news, a job that enabled her to care for her ailing mother until the elderly lady died in 1942. When the Royal Mail's South America service resumed in 1948, Violet couldn't resist signing on for two more years' service as a stewardess on the *Andes* sailing between Britain and Brazil. She finally retired in 1950, aged sixty-three, and moved to rural Suffolk, having spent a total of forty-two years of her life at sea, in which she had survived one collision, two sinkings and both World Wars.

In October 1946 the newly refurbished *Queen Elizabeth* made her belated maiden voyage as a passenger liner, and she carried the first four women officers ever to serve in a ship of the British merchant navy. Each one of the four 'lady assistant pursers' (LAPs) had been a Wren during wartime; one of them, Phyllis Davies, had made twenty wartime trips in the *Queen Elizabeth* as cipher officer. LAPs were involved in the berthing of passengers, making sure they were happy with their cabins, and moving them if necessary. On board the *Queen*s', they also provided a travel bureau, and could make hotel bookings, or arrange rail tickets for all over the States and boat trains for London or Paris. In addition, they looked after passengers' valuables in the safe deposits. As a reflection of their new officer status, the LAPs were allowed to watch films in the cinema and to dine in the passenger restaurant, and though they could choose from the first-class menu, they were not allowed lobster or caviar, and could only order smoked salmon once each voyage.

Despite the grip of austerity on post-war Britain, those passenger vessels that had survived the war, such as the *Queen Mary* and *Queen Elizabeth*, were once again linking Europe and North America on a weekly basis. They were particularly popular with royalty, millionaires and Hollywood stars. Though business was booming, passenger shipping was a competitive field, so international liner companies were keen to publicise any celebrities who favoured their ships over those of a rival.

After the war, the Duke and Duchess of Windsor frequently travelled between France and the USA on the most luxurious ships, with a peripatetic court of hangers-on and international demi-mondaines. They had titles, but no official role, beyond being in the vanguard of the so-called 'beautiful people', as Elsa Maxwell termed them. Elsa had always been a great collector of royalty and celebrities, and she had first met the duke in 1922, when he was Prince of Wales. However, her first encounter with Wallis was in New York in the winter of 1946, soon after the Windsors had moved into a suite above Elsa's in the Waldorf Astoria Tower Apartments. The couple had invited her for tea:

> I didn't know what to make of this tiny, rather ordinary woman who had climbed from a middle-class home in Baltimore to the threshold of Buckingham Palace. She has assurance and poise; her clothes were perfect. But I could detect none of the strong physical attraction she obviously held for the duke. Yet I did sense a terrific drive in the duchess, a drive comparable to the vibration you feel constantly on an ocean liner. The source or the nature of the force that made the duchess tick intrigued me more than ever.[7]

The Duke and Duchess of Windsor initially favoured Cunard ships. There is perhaps some irony that the man who refused to be king chose to travel on the dual flagships of the British merchant fleet. The first was named after his mother, Queen Mary, and the second after his sister-in-law, Queen Elizabeth. Neither woman ever forgave the duke, or Wallis, for the abdication. On the *Queen Mary* the Windsors regularly walked the 'measured mile' round the deck for exercise, accompanied by their pug. The duke frequented the bridge late at night to chat with the officers and smoke when Wallis was particularly snappy with him.

There was a travel boom in the late 1940s, and Cunard

benefited. The ocean liners were financially successful and contributed greatly to Britain's balance of trade in the late 1940s and early 1950s, because they brought in foreign income, especially dollars. By 1949 tourist berths on the *Queen Mary* and the *Queen Elizabeth* were booked up to a year in advance, and two months in advance for first class. Celebrity passengers flocked to travel the Atlantic and were much photographed, filmed and generally fêted by the shipping companies' PR departments, who had a symbiotic relationship with the press and media on both sides of the ocean. Readers of movie star magazines liked to know about the international travels of their idols, and Hollywood studios were anxious to plant stories about their employees' quirks and preferences, to promote forthcoming films. By the early 1950s Cunard had twelve liners in service to meet the demand. In high summer, on peak transatlantic sailings even the older ships such as *Mauretania* were packed to capacity. Of course, there were other shipping companies and competition was stiff, but there were some passengers who preferred to sail on specific ships and in their favourite cabins.

The success of passenger travel also benefited numerous land-based service industries that supplied the great ships, and many women's jobs ashore were dependent on the Atlantic Ferry. During the 1950s, whenever the *Saxonia II* docked in Liverpool between crossings, it unloaded 6,400 sheets, 8,500 table napkins and 16,000 towels to be washed in the giant laundries in the port. Housekeeping tended to be women's work too; the *Caronia II* would put in for its annual refurbishment every winter, and vast quantities of linen had to be cleaned, mended or replaced, along with 10,000 curtains, bedspreads and carpets, 4,000 pillows and 1,300 mattresses.

In post-war Britain rationing was gradually lifting but the age of austerity continued. The cities of Britain had been badly bombed, especially vital seaports such as the London Docks, Liverpool and Southampton. The workforce, particularly those who had served in the armed forces overseas, were keen to take jobs that allowed them to travel. Merchant navy jobs provided them with a decent income, and working on the big ships became even more attractive as a career option for young men and women. The opportunity to leave behind the bombsites and rubble, the poorly stocked shops, the dull and restricted food and the dismal British weather for the neon-lit, brilliantly coloured, smartly dressed and culturally vibrant cities of the USA and Canada had never seemed so appealing to the restless young. The fashions, style and music of the North American continent in the late 1940s and early 1950s, and the exuberant consumer culture on show there, were catnip to many British men and women. Commercial radio stations abounded in the States, playing popular music; in Britain the sole public broadcaster was BBC Radio, whose attitude to popular music was severely conservative. Television in the States was an everyday reality rather than an expensive novelty, with diverse channels competing to win consumers' attention. Fashions were bright and appealing, with new synthetic fabrics and glorious colours, far from the mud-coloured tweeds and muted twinsets worn by women of all ages back in Britain.

Crew members working on the transatlantic ships spent their spare time exploring the port cities, going to the movies, dancing, shopping. They used their pay and their tips to buy records, clothes, magazines and illustrated comics, and they often took commissions from friends and family back home who were desperate for novelty, a new dress, nylon stockings and dance music. Back home, they were known locally as the Cunard Yanks – well-dressed, affluent and snappy dancers, they were part of the burgeoning youth culture, especially in Liverpool, a city always open to new ideas, and perennially ready for a party. Jazz and black music records found their way into clubs in Liverpool and London, permeating the music scene of the 1950s.

With the return of peace and a period of stability and regrowth, there were more opportunities for women who wanted to work at sea. Former Wrens were demobbed, and their wartime service in support of the Royal Navy made them well-suited to life afloat in the merchant navy, as they had appropriate knowledge, as well as 'people skills'. Some former Wrens sought jobs as stewardesses or cooks on ships. Fourteen of them joined the combined cargo and passenger ship *La Cordillera*, signing on for a five-month round-the-world trip. Significantly, their appointment sparked a special enquiry by the National Union of Seamen, to consider the employment of women instead of men aboard ships. The women were allowed to become union members, and each earned £20 a month.

While sea transport remained the only viable way for passengers to cross the Atlantic during the late 1940s and early 1950s, it was apparent that air travel would eventually become a practical reality for those who needed to travel, and diversification was necessary for any shipping company determined to survive. One solution was the revival of leisure cruising. As cultural historian and veteran cruise-ship lecturer Paul Atterbury observed, 'The ship itself was a destination.'[8] For those who could afford a 'holiday afloat' the ship provided all the comforts of staying in a hotel while being conveyed to exciting places, 'and you only have to unpack once!'. The cruise ship offered swimming pools, ships, bars, restaurants, games, exercise, relaxation, libraries, evening entertainments, unlimited food and drink, with a level of service now largely unknown on shore. There was generally a carnival atmosphere too, a sense of hedonism and unbridled fun. Leisure cruising could now be marketed as an escape, a holiday afloat, an end in itself.

Of course, cruising as a holiday, let alone as a way of life, wasn't for everyone. Writer and war correspondent Martha Gellhorn, an inveterate traveller, constantly impatient to get to her next destination, loathed even the idea of leisure cruising: 'It bores me to even think of such a trip, not that I mind luxury and lashings of delicious food, and starting to drink at eleven o'clock with a glass of champagne to steady the stomach. But how about the organised jollity, the awful intimacy of tablemates, the endless walking round and round because you can't walk anywhere else, the claustrophobia?'[9]

In the 1950s for many cosmopolitan travellers the advent of long-distance commercial travel by air was revolutionary. The prospect of air travel between continents had intrigued the public for decades before it became a reality, and it particularly concerned those whose business was mass transport on a global scale. As early as summer 1914, the *Aquitania*'s on-board newspaper reported on Monsieur Blériot, the French pioneer of aviation, and his visit to his aeroplane factory at Brooklands in England. The same paper carried an interview with Count Zeppelin, describing his plans for airships that might cross the Atlantic. Both men predicted that an Atlantic crossing by some form of aircraft, whether by aeroplane or some sort of dirigible, was likely by mid-1915. But it wasn't until after the Second World War, with the development of jet aircraft, which greatly improved the fuel efficiency of new aeroplanes, that the range of planes was expanded so as to make commercial transatlantic flight possible. New materials developed during the war years, such as aluminium, made possible the construction of lighter, larger aircraft, capable of carrying greater numbers of passengers. In 1953 the De Havilland Comet became the first commercial jet airliner, and within two years a number of airlines such as Pan American Airlines and Air France were ordering passenger jets, such as the Boeing 707 and the Douglas DC-8. Long-distance travel was now a practical and affordable alternative to ocean voyages, and in 1957, for the first time, more people crossed the North Atlantic by air than by sea. As a means of long-distance transportation the great ships were finally – and fatally – overtaken by long-haul aircraft.

There were distinct advantages to transatlantic travel by air: the costs were comparable, and the Boeing 707 made it possible to fly the Atlantic in under eight hours, compared with five days afloat on even the fastest ocean liner. The flights were comfortable and flying at such high altitudes lessened the experience of turbulence. The planes' interiors were modern and glamorous, and the on-board flight attendants, the 'air stewardesses', were young, slim, beautiful and often dressed in futuristic-looking fashions. In short, air travel was progressive and in keeping with the more optimistic Brave New World scientists and politicians had promised after the war. The rich took up so-called airliners and flying became the transport choice of the international jet set.

The *Queen Mary* made its thousandth Atlantic crossing on 25 September 1957. By the middle of 1959 two-thirds of transatlantic passengers between Britain and America were travelling by jet, especially in the off-peak winter months when sailing the Atlantic was an unappetising prospect. In

December 1960 the Cunard passenger ship *Parthia* (a vessel only thirteen years old and the favourite ship of Katharine Hepburn) sailed from Liverpool to New York with just twenty-five passengers occupying its 251 berths. By the early 1960s, 95 per cent of passenger traffic across the Atlantic was by aircraft, and this effectively marked the end of the ocean liners as a form of mass transportation. However, the shipping companies adjusted to their changing circumstances, and expanded their role as cruise ship operators, providing luxurious 'floating hotels', the forerunners of the international cruising industry that continues in the twenty-first century.

Conclusion: Sailing into the Sunset

Millions of women's lives were profoundly changed by the phenomenon of transatlantic travel by ship in the first half of the twentieth century. The ocean liners had a huge impact on the economies and geopolitics of powerful nations, but they also transformed the life chances and livelihoods of individuals from all nations, classes and backgrounds, by providing them with unparalleled opportunities for independence, adventure and travel.

Sea transport was a practical necessity for anyone who wished to travel between the great land masses of Europe and North America, and ocean-going vessels of all sizes were the only practical option until the middle of the twentieth century. However, there were also psychological aspects of sea voyages the made a transatlantic trip a memorable experience, at the very least a welcome novelty, if not a great adventure, for most individual travellers. Through the growth of the popular media, especially in the years between the wars, the general public came to associate technologically sophisticated ships, each bigger and better than the last, with their own optimistic notions of progress and modernity. National pride was invested in and exemplified by these great enterprises, and the vessels were described lyrically, almost poetically, by the commentators of the day. During the 1920s and 1930s, the Ocean Greyhounds were the mightiest man-made moving objects in existence, commanding awe and admiration in equal measure, as they loomed into view at the end of piers and docks.

For the passengers, the voyage itself provided a unique and slightly surreal experience; there was the necessary duration of the trip, at least five days, often seven days or more. The strictures and routines of normal life were suspended, and replaced with a hectic social whirl with one's fellow travellers, accompanied by music, dancing, entertainment, sports, gambling and the ever-present possibility of romance or intrigue. In a more formal era, the ship's transient population felt themselves to be 'on show' when in public, and dressed and behaved accordingly. The theatricality of the setting was acknowledged, even celebrated, by the ship's designers, whose interiors ranged from historical pastiche to art deco grandiosity, complete with mirrors and elegant staircases, all the better to effect the *grande descente*. Even the measure of time itself was flexible, with the ship's clocks advancing or receding each day. In addition, there was the lurking knowledge that an iceberg or a storm could imperil the ship, which added a scintillating frisson of danger.

The golden age of transatlantic travel, the two decades between the wars, coincided with huge social changes for women, and the great ships were often their means of escape, from old lives to newer, hopefully better ones. The Great War had propelled many of them into the workplace, where they had taken on roles previously occupied by men. After the Armistice, many women were reluctant to give up their new jobs to accommodate the menfolk returning from war. Financial and personal independence of a sort had been hard-won – now it became a badge of honour, along with the vote. The massive death toll meant there were hundreds of thousands of 'surplus women' in Britain who could not expect to marry, and who needed to earn their own living, whether in new professions, or by travelling overseas in search of better opportunities and bigger gene pools. For women passengers, getting on to a ship and crossing the Atlantic was a professional and personal gamble. The voyage itself was emblematic of women acting with agency, even if it required a leap of faith. By travelling they could reinvent themselves, discarding their old existences to forge new ones. Hedy Lamarr left behind her husband and family, her privileged if restricted life in Austria, and even her name, to become a Hollywood star and a pioneering scientific inventor. Tallulah Bankhead took London by storm, using her acting abilities and her wits to delight and thrill the Bright Young Things of 1920s London, while outraging the British establishment. Josephine Baker became the toast of liberal Paris, celebrated for her extraordinary talents, rather than restricted and discriminated against because of the colour of her skin.

Some travelled regularly on the Atlantic Ferry as a necessary part of their lives, constantly linking aspects of their individual careers between continents. Hard-working professional performers such as Lady Diana Cooper and Adele Astaire, businesswomen such as Sibyl Colefax, and pioneering politicians such as American-born British MP Lady Astor relied on the great ships to help them pursue their dreams.

Events during a voyage could dramatically affect the subsequent course of women's lives. Lady Lucy Duff-Gordon and her husband, who had previously enjoyed lives of privilege, luxury and social status, escaped the sinking of the *Titanic*, but they could not avoid the malicious rumours and subsequent damage to their public reputations. Thelma Furness's sea-borne love affair with Aly Khan wrecked her long-standing relationship with the Prince of Wales, and sparked the romance between him and Wallis Simpson, leading to the abdication.

Individuals' decisions to emigrate had life-changing consequences, particularly for those travelling in third class. The Riffelmacher family exchanged the poverty and hyper-inflation of 1920s Germany for the prosperity and security of the American mid-West. Mary Anne MacLeod set out in 1930 as an eighteen-year-old domestic servant from an impoverished Scottish island and became the mother of the President of the United States.

Celebrities often saw an ocean voyage as a way to promote their careers. Marlene Dietrich and Elsa Maxwell were always willing to pose for photos, recognising the benefit of good publicity. Others – reclusive individuals such as millionairess Barbara Hutton and publicity-shy screen diva Greta Garbo – craved the seclusion and privacy of a transatlantic ship. Some saw the journey itself as a valuable experience, which provided good 'material' for their careers. Inter-war writers such as E.M. Delafield and Anita Loos responded to the Zeitgeist, setting their popular fiction largely aboard the inter-war transatlantic liners. By contrast, Martha Gellhorn, battle-hardened war correspondent, found her way on to whatever vessel would take her across the sea to get to the story, regardless of the personal risk.

Some unscrupulous women embarked on ocean voyages for their own nefarious purposes: 'ocean vamps' and black-mailers found the big ships a happy hunting ground, while female accomplices of card-sharps would act as lookouts or spies for their partners. In addition, for both passengers and crew, there was money to be made smuggling contraband, or illicit alcohol during the febrile days of Prohibition.

For women in steerage or third class, transatlantic travel was often an endurance test, a challenging and uncomfortable transition between the familiar Old World and the completely unknown. Some, like hard-pressed former stewardess Christiana Klingemann, were so desperate to escape the economic hardship crippling their home countries that they were forced to stow away. Others, like Earnestine Tennenbaum, saw the New World as a refuge from the rising tide of pogroms, purges and anti-Semitism that threatened to engulf them, and viewed the ship that transported them and their families as a magic carpet to freedom.

In wartime the ship could be both a means of escape and at the same time an extraordinarily vulnerable target, alone and unprotected on an open sea. It took courage and nerve to cross the Atlantic, knowing that one's ship might be torpedoed, or to volunteer to escort 'seavacs', the children being sent abroad by ship for their safety. After the Second World War, troops were repatriated, families were reunited and GI brides with their young offspring bravely set out for unknown countries, joining husbands they often barely knew at the other end of long ocean voyages. Their journeys were made more tolerable by the ministrations of Red Cross nurses and stewardesses.

The return of peace in the late 1940s and early 1950s attracted film stars and royalty, who travelled on hastily refurbished ships, where unparalleled levels of service and luxury prevailed, recalling almost a nostalgic era. However, the same 'beautiful people' were quick to take up inter-continental air travel when it became an affordable and viable way of crossing the Atlantic. Passenger ships were either sold off or transformed into cruise ships, to meet the new leisure market. Shipping companies expanded into leisure cruising, catering for wealthy individuals such as Clare L. MacBeth, the 'perpetual passenger' who spent fourteen years living aboard the *Caronia*, and privilege and pleasure were their new maxims.

During the golden age of ocean-going ships passengers relied on the expertise, advice and practical assistance of working women, those who had chosen the sea as their careers. The sea-going roles available to women increased greatly during the first five decades of the twentieth century. Initially they were employed, somewhat grudgingly, only in traditional, 'nurturing' roles, as matrons, to look after the sick, and then as stewardesses, providing comfort and practical assistance to their female and junior clients. Outstanding characters, such as Violet Jessop, the Unsinkable Stewardess, served for decades on ships because they enjoyed the working life afloat despite its many tribulations, discomforts and dangers. Nursery nurses and nursing sisters augmented the stewardesses' ranks in the 1920s, in response to a boom in both tourism and mass migration. Conductresses such as Edith Sowerbutts had considerable status within the ship's hierarchy, as they were employed on transatlantic routes to guard the moral welfare of the unaccompanied, and those deemed most vulnerable to sex traffickers. As a result they often clashed with male authority figures on board the ships, which did not make them popular, but did engender respect.

Between the wars, a woman's place was no longer necessarily 'in the home' (whether her own, or cleaning the silver in someone else's). Seafaring women tended to be self-sufficient and independent characters. Their work was physically demanding, their working environment might be physically cramped and occasionally dangerous, and their workmates could be unpleasant. But they challenged popular assumptions of what women's capabilities and aptitudes were believed to be in the mostly macho world of the merchant navy, by proving themselves to be resourceful, intelligent and brave. Those with individual and unique talents, such as swimming instructresses like Hilda James, were initially taken on to meet women's changing perceptions of sports and leisure, but soon tackled extra responsibilities and were the forerunners of women working in the leisure cruising industry today. Hairdressers such as Ann Runcie, beauticians, masseuses and seamstresses were employed to enhance the appearance of their customers, and were well-rewarded with salaries and tips. By the 1930s stenographers were making inroads into the holy of holies, the purser's office, and were entrusted with junior clerking roles, though they always remained subordinate to male officers until after the Second World War.

The economies of British cities such as Liverpool and Southampton were intimately bound up with the fortunes of the merchant shipping fleet, with subsidiary industries such as huge laundries, victualling and maintenance employing numerous women around the ports. Many seafaring women had grown up in families where generations of menfolk had served at sea, and their incomes were essential to support households with salaries and tips earned on the lucrative North Atlantic run, especially in the inter-war years. Whenever there was an economic downturn, female seafarers would be laid off, and forced to seek other work ashore, though they often returned to the ships when demand for their services increased. During the two World Wars, stalwart seafaring women such as Maida Nixson often acquitted themselves well while afloat, remaining calm under pressure, even when under enemy attack. There were also those individuals who displayed extraordinary courage, such as Victoria Drummond MBE, the ship's engineer, an exceptional pioneer in every way.

After the Second World War, the merchant navy recruited former Wrens, women who had joined the armed forces but who had previously only gone to sea in restricted circumstances. For the first time some were deemed sufficiently experienced and motivated to work their way into officer roles as LAPs, later becoming pursers in their own rights. By the late 1940s there were many new opportunities for working women seafarers, who needed to earn money but were also keen to exchange the gritty, grey austerity of war-torn Britain for the chance to travel. Working women at all levels subsequently found useful and worthwhile careers afloat in the fast-expanding seaborne leisure industry, and in time their presence on board cruise ships was accepted and largely respected by their male colleagues.

By the beginning of the twenty-first century, a small number of determined women had finally made it on to the bridge, that bastion of male control, as deck officers. While 98 per cent of the workforce involved in the shipping industry worldwide is still male, the leisure and cruising sector offers much wider representation. Nowadays nearly 20 per cent of the global cruise industry employees are female, and in some particularly enlightened companies, up to 22 per cent of the officers on board are women. Cunard appointed its first woman as captain of a cruise liner in 2010. Inger Klein Thorhauge had first worked as a stewardess in her school holidays, aged sixteen, and came to love the life afloat. She obtained her Master's Licence in 1994, joined Cunard in 1997 and worked her way up the ranks. In July 2016 Captain Inger, aged forty-three, was at the helm of the giant cruise ship

Queen Elizabeth as it sailed into Liverpool to celebrate the centenary of the iconic Cunard Building.

Edith Sowerbutts, feisty veteran of the transatlantic run, died in 1992, but she would have been delighted to have witnessed such an important symbolic achievement for a woman seafarer, to be in charge of a mighty ship. Presciently, in her memoirs she had written: 'Nobody had visualised that female staff, other than the very necessary stewardesses, would ever be carried on ocean-going liners ... We, of my generation, comprised the thin end of the wedge. Women would eventually be signed on for seagoing positions once considered to be male preserves.'[1]

Those pioneering and intrepid women who sailed the Atlantic during the golden age of travel, whether as passengers or seafarers, had their lives transformed by their experiences, mostly for the better. Their motivations were as diverse as their personalities, but for each of them, to embark upon a sea voyage at all was to take a step into the unknown. Every life is precious and unique to the individual living it; the vast majority of women whose lives are now largely unknown or unrecorded, as well as those who became famous and celebrated – or notorious – in their lifetimes, were willing to sail the ocean in hope, rather than stay ashore in fear. In the words of Grace Brewster Murray Hopper, United States Rear Admiral and pioneering computer scientist: 'A ship in harbour is safe. But that is not what ships are built for.'[2]

Ships' Names, by 'Lucio'

I think that few would care to sail
On board the barque Dyspeptic,
Though there's something rather fit and hale
About the Antiseptic;
But many a sport of the frugal sort
Would book by the Economic,
And of honest mirth there should be no dearth
Aboard the good ship Comic.
The millionaire would pay his fare
For a passage per Aurific
And the bishop's scope would be met, I hope,
By a berth on the Beatific;
The countryman would feel at home
On board the bold Bucolic
And the chorus girl could cross the foam
In fine style on the Frolic.
Geologists could much enjoy
A jaunt on the Jurassic,
And the doctors one might well decoy
With the Clinic or Boracic;
All politicians outward bound
Would sail by the Polemic,
And the professorial folk be found
Aboard the Academic.
The chemist's tip would be, wait for the ship

That bears the name Synthetic.
The maid of today would sail away
On the Lipstic(k) or Cosmetic;
SS Hydraulic Drys* might find
Sufficiently symbolic,
While the Wets,[1] *of course, would remain behind*
And wait for the Alcoholic.

(*White Star Line Magazine*,
June 1927, reproduced
from the *Manchester Guardian*)

[1] *This was written in the era of Prohibition.*

Acknowledgements

So many people were very helpful in the research for this book, pooling their knowledge and offering valuable support. I would particularly like to thank the Special Collections staff at the University of Liverpool, who look after the Cunard Archives; they were great sources of advice and expertise, especially Robyn Orr, Siân Wilks and Elizabeth Williams. Roddy Murray, Founding Director of An Lanntair in Stornoway was very generous with his knowledge, and I am similarly indebted to the patience and expertise of many staff in the archives section of the Imperial War Museum, the reading rooms of the British Library, and those who maintain the Women's Library at the London School of Economics.

Lisa Highton, Kate Craigie and Katherine Burdon at Two Roads deserve to be thanked for their vision and their patience, along with Diane Banks, Martin Redfern, Charles Spicer, and Morag Lyall. Grateful thanks are due to my family, especially Sarah Evans, David Meurig Evans and Martin Roberts. Friends who provided insights, leads and general encouragement are too numerous to mention, but special thanks are due to Jacq Barber, Mike Calnan, Lauren Taylor, Harvey Edgington, Sarah Payne and Sarah Holloway. Steve Price and Mark Fifield provided much practical help and information, while Phillip Arnold and Philip Baldwin were amusingly informative about the culture and protocols of life on board cruise ships. Thank you all.

Picture Credits

Alamy Stock Photo: 1,2 above, 3, 5 above, 6 below, 8 above left and centre, 10 above and below, 11, 12 above, 13 above, 15. Bridgeman Images: 5 below/photo © Tallandier. Crown Copyright/Public Domain: 2 centre right. Getty Images: 4 above, 6 above, 12 below, 14 above left and centre, 16 below. Imperial War Museum London: 7 above (Private Papers of Miss E.F.M. Sowerbutts. Catalogue: Documents 107), 7 below (A 7842a), 16 above (NA 11895). Library of Congress: 4 below (LC-B2-5708-7 P&P), 9 below (LC-B2-5950-11 P&P). Mirrorpix: 9 above. *Ring Twice for the Stewardess*, Maida M Nixson, John Long Ltd (London), 1954: 14 below left. Shutterstock.com: 8 below left/AP, 13 below/Underwood Archives. By courtesy of The University of Liverpool Library, Cunard Archive: 2 below left (D.785/3). Captain Stephen Gronow: v (D42/PR2/2/47/2).

Notes

Prologue

1 P.G. Wodehouse, *The Girl on the Boat*, George H. Doran 1922.

Introduction

1 Charles Dickens, *American Notes*, 1842.
2 Ibid.
3 *The Times*, 24 August 1859.
4 Mrs H. Coleman Davidson, *What Our Daughters Can Do for Themselves: A Handbook of Women's Employments*, Smith, Elder, 1894, pp. 269–70.
5 'The Economic and Social Value of Immigrant Women', *Atlantic Monthly*, September 1907.

Chapter 1: Floating Palaces and the 'Unsinkable' Violet Jessop

1 Violet Jessop, *Titanic Survivor*, ed. John Maxtone-Graham, Chapman & Hall 1998, p.49.
2 Ibid., p. 58.
3 Joseph Ismay, *United States Senate Inquiry: Day 1* [Transcript]. Retrieved from https://www.titanicinquiry.org/ (original interview held April 19, 1912).

4 Osbert Sitwell, *Great Morning*, 1948, Macmillan, London p. 259.

5 Anon., *The Great Cunarder RMS Aquitania*, 1914.

6 Arthur Davis, RIBA interview, 1922, reproduced in Gérard Piouffre, *First Class*, p. 118.

Chapter 2: From the Ritz to the Armistice

1 Violet Jessop, (ed.) John Maxtone-Graham, *Titanic Survivor*, p.190.

2 *Country Lfe*, 15 May 1915, p. 649.

3 Typewritten letter from Cunard Chairman A. A. Booth to Charles P. Sumner, 8 May 1915. Cunard Archives, University of Liverpool SCA.

4 Henry Eaves Papers, 'The Cunard Steam Ship Co. Ltd, 1840-1930', unpublished manuscript in the Cunard Archives, University of Liverpool SCA, p. 32.

5 Quoted in John P. Eaton and Charles A. Haas Ltd., *Falling Star: Misadventures of White Star Line Ships*. p. 177.

6 *Daily Graphic*, 18 June 1918.

7 Jessop, op.cit. p. 228.

8 'Our Ladies Corner', *Cunard Line*, November 1918, p. 19.

Chapter 3: Sail Away: Post-war Migration and the Escape from Poverty

1 'Mothers of the Sea', *Cunard Line*, April 1921, p. 134.

2 *Cunard Line*, November 1919, p. 29.

3 'Have an Ambition', *Cunard Line*, August 1920, p. 82.

4 Cunard Archives, Liverpool. Accessed 19 February 2018. D.785/1 – 'Emigrations on Tyrrhenia and Laconia': Typewritten account of the early years of Marie Riffelmacher/

Ruffelmacher, written up by her granddaughter Kathleen Stroemer Czuba.

5 Frederick A. Wallis. What about the Immigrant? *The Rotarian*, 1921, p. 249.

Chapter 4: The Roaring Twenties

1 Charles T. Spedding, *Reminiscences of Transatlantic Travellers*, T. Fisher Unwin, 1926, p. 169.

2 Basil Woon, *The Frantic Atlantic: An Intimate Guide to the Well-Known Deep*, Alfred A. Knopf, 1927, p. 199.

3 Violet Jessop, *Titanic Survivor*, p. 226.

4 Gina Kaus, *Luxury Liner*, Cassell, p. 210.

5 Alfred Duff Cooper, *Old Men Forget: The Autobiography of Duff Cooper*, Rupert Hart-Davis, p.120.

6 Christopher Sykes, *Nancy: The Life of Lady Astor*, Panther Books, 1979, p. 280.

7 Spedding, op.cit., p. 269.

8 Jessop, op.cit., p. 229.

9 S.E. Lasher, 'Travel Fashions seen in the Shops', *Cunarder*, May 1921, p. 43.

10 Doris Estcourt, 'Homeward Bound', *White Star Magazine*, July 1926, p. 308.

11 Humphrey Carpenter, *Robert Runcie*, Hodder & Stoughton, 1977, p. 55.

12 *White Star Magazine*, July 1930, p. 379.

Chapter 5: Edith and Her Contemporaries

1 Edith Sowerbutts, *Memoirs of a British Seaman*, unpublished manuscript, IWM, p. 78.

2 Ibid., p. 15.

3 *White Star Magazine*, October 1926, pp. 34–5.

4 Sowerbutts, op.cit., p. 99.

5 Ibid., p. 53.

6 *Cunard Line*, May 1921, p. 162.

7 Sowerbutts, op.cit., p. 85.

8 Ibid., p. 78.

9 *Cunard Line*, October 1920, p. 111.

10 *Manchester Guardian*, 9 March 1928.

11 *Westminster Gazette*, 7 November 1924.

12 'The Captainess', *Daily Chronicle*, 12 March 1926.

13 *Daily Chronicle*, 26 October 1925.

14 *Evening Standard*, 4 November 1927.

15 *White Star Magazine*, September 1927, p. 10.

16 *Westminster Gazette*, 3 October 1927.

17 *Manchester Guardian*, 9 January 1929.

18 Sowerbutts, op.cit., p. 73.

19 Sowerbutts, op.cit., p. 72.

Chapter 6: For Leisure and Pleasure

1 Edith Sowerbutts, *Memoirs of a British Seaman*, p. 87.

2 Charles T. Spedding, *Reminiscences of Transatlantic Travellers*, p. 278.

3 Basil Woon, *The Frantic Atlantic*, pp. 144–5.

4 Stella Margetson, *The Long Party: High Society in the Twenties and Thirties*, Gordon Cremonesi, 1974, p. 123.

5 Richard Collier, *The Rainbow People : A Gaudy World of the Very Rich and Those Who Served Them*, Dodd Mead; 1st edition 1984, p. 73.

6 Woon, op.cit., p. 68

7 Elspeth Wills, *Stars Aboard: Celebrities of Yesteryear Who*

Travelled Cunard Line During the Golden Age of Transatlantic Travel, London: Open Agency, 2003, p. 5.

8 Evelyn Waugh, 'The Gentle Art of Being Interviewed', *Vogue*, July 1948.

9 Woon, op.cit., p. 3.

Chapter 7: Depression and Determination

1 Edith Sowerbutts, *Memoirs of a British Seaman*, p 113.

2 Gloria Vanderbilt and Lady Thelma Furness, *Double Exposure*, Frederick Muller, 1959, p. 269.

3 Ibid., p. 298.

4 Ibid., p. 328.

5 Sowerbutts, op.cit., p. 120.

Chapter 8: The Slide to War

1 *The Queen Mary: Greatest Ocean Liner*, BBC 2, first shown 24 May 2016, STV Productions for BBC 2

2 Meghan L. McCluskey, 'Interview with Ann Davis Thomas', *Interior Design: Student Creative Activity*. Paper 1 http://digitalcommons.unl.edu/archidstuca/1 University of Nebraska, 29.11.2004.

3 Cecil Beaton, 'Reviewing the Queen', *Vogue* (US edition), 1 July 1936.

4 Edith M. Vigers, *Evening Standard*, 11 February 1936.

5 Henry Eaves, *The Cunard Steam Ship Co Ltd.*, 1840–1930, unpublished manuscript, University of Liverpool SCA, p. 436.

6 'German Woman as Sea Captain', *Daily Telegraph*, 13 April 1937.

7 Patt Morrison, 'A Royal Mess But One Heck of a Story', *The Los Angeles Times*, 17 October 1999.

8 Richard Collier, *The Rainbow People*, p. 169.

9 Richard Rhodes, *Hedy's Folly: The Life and Breakthrough Inventions of Hedy Lamarr, the Most Beautiful Woman in the World*, Vintage Books, 2011, p. 108.

10 Sowerbutts, *Memoirs of a British Seaman*, p. 177.

11 Violet Jessop, *Titanic Survivor*, p. 21.

12 Sowerbutts, op.cit., p. 210.

13 'Lily Pons sings as gale rocks giant liner', *Evening News*, 11 April 1938

14 *White Star Magazine*, May 1927, p. 224.

15 *The Queen Mary: Greatest Ocean Liner*, BBC 2

16 Ashley Halsey III, 'The Queen Mary Saved Hundreds of Jews from the Nazis, even as St Louis was Turned Away', *Washington Post*, 21 September 2019.

17 Sowerbutts, op.cit., p. 198.

18 Dorothy Scobie, *A Stewardess Rings a Bell*, Stylus, 1990, p. 84.

Chapter 9: Women under Fire

1 Edith Sowerbutts, *Memoirs of a British Seaman*, p.198

2 Ibid., p. 213.

3 Ibid., p. 4.

4 Maida Nixson, *Ring Twice for the Stewardess*, John Long, 1954 p. 38.

5 Ibid., p. 130.

6 Sowerbutts, *Memoirs of a British Seaman*, p. 12.

7 Ibid., p. 16.

8 Nancy Cunard, untitled manuscript in the form of a journal, kept from 31 July to 21 August 1941. The Harry Ransom Humanities Research Center Library, University of Texas at Austin.

9 Ibid.

10 'Woman Ship's Officer: She Escaped Bombs and Torpedoes' *Evening Standard*, LSE Archives, June 1940.

11 Anon., 'A Woman on the High Seas', unattributed article in *Woman Engineer* magazine, 1941.

12 Sowerbutts, op.cit., p. 42.

13 'Forgotten Women of the Sea', *Daily Herald*, 4 June 1942.

14 'Women in Shipbuilding', *The Times*, 16 June 1943.

15 Martha Gellhorn, 'The First Hospital Ship', in *The Face of War*, Granta Books, 1998, p. 119.

16 Nixson, op.cit., p. 149.

17 'Cunard Pays Tribute to War Service", unattributed article in *The Maritime Executive* 24.5.2015

Chapter 10: Romance, Repatriation and Recovery

1 Sylvia Duncan and Peter Duncan, *The Sea My Steed*, Robert Hale, 1960, p. 14.

2 Interview, Southampton City Council Oral History Unit.

3 Maida Nixson, *Ring Twice for the Stewardess*, p. 171.

4 Ibid., p. 174.

5 Duncan and Duncan, op.cit., p. 150.

6 *New York Times*, 17 December 1949.

7 Elsa Maxwell, *Elsa Maxwell's Own Story*, Little, Brown, 1954, pp. 299–300.

8 *The Golden Age of Liners*, BBC 4. Presented by Paul Atterbury and made for the Timeshift series, first broadcast on BBC 4 on 22 October 2009.

9 Martha Gellhorn, *Travels with Myself and Another*, p. 286.

Conclusion

1 Edith Sowerbutts, *Memoirs of a British Seaman*, p. 5.
2 Grace Brewster Murray Hopper, Address given at Trinity College, Washington, USA. Reported in *Time*, 22 June 1987.

Bibliography

Anon., *The Great Cunarder RMS Aquitania: The World's Wonder Ship*, Cunard Steam Ship Company, 1914

Anon, 'The Quadruple-screw Turbine-driven Cunard liner "Aquitania"', *Engineering* [magazine], 1914

Anon, *Cunard on War Service*, Cunard Line, 1919

Anon, *The Story of a Great Liner*, Royal Mail Steam Packet Company, 1924

Baker, Paul and Stanley, Jo, 'Hello Sailor! The Hidden History of Gay Life at Sea', Routledge, 2014 (First pub. Pearson Education, 2003)

Banstead, C.R., *Atlantic Ferry*, Methuen, 1936

Barrow, Andrew, *Gossip: 1920–1970*, Hamish Hamilton, 1978

Benton, Charlotte, Benton, Tim and Wood, Ghislaine, *Art Deco 1910–1939*, V&A Publications, 2003

Blair, Gwenda, *The Trumps: Three Generations that Built an Empire*, Simon and Schuster, 2001

Braynard, Frank O., *Picture History of the Normandie*, Dover Publications, 1987

Brendon, Piers, *Thomas Cook: 150 Years of Popular Tourism*, Secker and Warburg, 1991

Brinnin, John Malcolm, *The Sway of the Grand Saloon: A Social History of the North Atlantic*, Barnes & Noble, 1971 and 1986

Brinnin, John Malcolm and Gaulin, Kenneth, *Grand Luxe: The Transatlantic Style*, Bloomsbury, 1988

Buckle, Richard (ed.), *Self Portrait with Friends: The Selected Diaries of Cecil Beaton, 1926–1974*, Weidenfeld and Nicolson, 1979

Burns, C. Delisle, *Leisure in the Modern World*, George Allen and Unwin, 1932

Carey, Gary, *Anita Loos: A Biography*, Alfred A. Knopf, 1988

Carpenter, Humphrey, *Robert Runcie: The Reluctant Archbishop*, Hodder and Stoughton, 1997

Chirnside, Mark, *RMS Aquitania: The Ship Beautiful*, History Press, 2008

Claridge, Laura, *Emily Post: Daughter of the Gilded Age, Mistress of American Manners*, Random House, 2009

Cooke, Alastair, *America*, BBC Publications, 1973

Coons, Lorraine and Varias, Alexander, *Tourist Third Cabin: Steamship Travel in the Inter-War Years*, Amberley Publishing, 2016

Cooper, Diana, *Darling Monster: The Letters of Lady Diana Cooper to her son John Julius Norwich, 1939–1952*, Chatto and Windus, 2013

Coward, Noël, *Autobiography*, Methuen, 1986

Cunard Daily Bulletin: Aquitania, 1914 (On-board daily newspaper produced for passengers travelling the Atlantic on *Aquitania*)

Cunard Line, 1918–25 (Staff magazine)

Czube, Kathleen Stromer, *Emigrations on Tyrrhenia and Laconia. Typewritten Account of the Early Years of Marie Riffelmacher/Ruffelmacher, Dictated to and written up by her Grand-daughter Kathleen Stroemer Czuba*, Cunard Archives

Davenport-Hines, Richard, *Titanic Lives: Migrants and Millionaires, Conmen and Crew*, Harper Press, 2012

Dawes, Frank Victor, *Not in Front of the Servants*, rev edn, Hutchinson, 1984 (First pub. Hutchinson, 1973)

Day, Susan, *Art Deco and Modernist Carpets*, Thames and Hudson, 2002

Delafield, E.M., *The Provincial Lady in America: A Compendium Edition of Collection Novels*, Amazon, n.d.

Dempsey, John, *I've Seen Them All Naked*, Alden Press, Oxford, 1992

Desert Island Discs, Tallulah Bankhead, interviewed by Roy Plomley, broadcast on BBC Home Service, 14 December 1964

Donaldson, Frances, *P.G. Wodehouse: The Authorized Biography*, Futura, 1983

Donzel, Catherine, *Luxury Liners: Life on Board*, Vendome Press, 2006

Duff Cooper, Alfred, *Old Men Forget: The Autobiography of Duff Cooper*, Rupert Hart-Davis, 1953

Duff-Gordon, Lucy, *Discretions and Indiscretions: Edwardian Couturier, It Girl and Titanic Survivor*, Spitfire Publications, 2019 (First pub. Frederick A. Stokes, 1932)

Duncan, Sylvia and Duncan, Peter, *The Sea My Steed: The Personal Story of Captain Donald Sorrell*, Robert Hale, 1960

Eaton, John P. and Haas, Charles A., *Falling Star: Misadventures of White Star Line Ships*, Patrick Stephens, 1990

Edington, Sarah, *The Captain's Table: Life and Dining on the Great Ocean Liners*, Conway, 2011 (First pub. National Maritime Museum Publishing, 2005)

Eliot, Marc, *Cary Grant: A Biography*, Three Rivers Press, 2004

Etherington-Smith, Meredith and Pilcher, Jeremy, *The 'It' Girls*, Hamish Hamilton, 1986

Feifer, Maxine, *Going Places: The Ways of the Tourist from Imperial Rome to the Present Day*, Macmillan, 1985

Finamore, Daniel and Wood, Ghislaine (eds.), *Ocean Liners: Glamour, Speed and Style*, V&A Publishing, 2017

Flanner, Janet, *Paris was Yesterday: 1925–1939*, Virago, 2003

Fodor, Eugene (ed.), *1936 … On the Continent*, facsimile edn of original 1936 pub., Fodor's Travel Guides, 1985

Fort, Adrian, *Nancy: The Story of Nancy Astor*, Vintage Books, 2013

Fox, Stephen, *Transatlantic: Samuel Cunard, Isambard Brunel, and the Great Atlantic Steamships*, HarperCollins, 2003

Freeman, Roydon, *Sea Travel*, St Catherine Press, 1930

Fussell, Paul, *Abroad: British Literary Travelling Between the Wars*, Oxford University Press, 1982

Gardiner, Juliet, *The Thirties: An Intimate History*, HarperPress, 2011

Gellhorn, Martha, *The Face of War*, Granta Books, 1998

Gellhorn, Martha, *Travels with Myself and Another: Five Journeys from Hell*, Eland, 2002

Good Housekeeping Institute, *A War Bride's Guide to the USA* (1945), republished by Collins & Brown / Anova, 2006

Gordon, Lois, *Nancy Cunard: Heiress, Muse, Political Idealist*, Columbia University Press, 2007

Hadfield, John (ed.), *Cowardy Custard: The World of Noël Coward*, Heinemann, 1973

Halsey III, Ashley, 'The Queen Mary Saved Hundreds of Jews from the Nazis, even as St Louis was Turned Away', *Washington Post*, 21 September 2019

Hemingway, Ernest, *A Moveable Feast*, Arrow Books, 2011 (First pub. 1964)

Hoare, Philip, *Noël Coward: A Biography*, Sinclair-Stevenson, 1995

Hunter-Cox, Jane, *Ocean Pictures: The Golden Age of Transatlantic Travel, 1936–1959*, Webb and Bower, 1989

Hurd, Archibald, *A Merchant Fleet at War*, Cassell,1920 (Book sourced by Sian Hicks, Cunard Archivist at Liverpool University SCA, reference R/HE 945.C9.H95) Consulted February 2017

Hurd, Archibald, *History of the Great War: The Merchant Navy*, Vol. 3, Spring 1917 to November 1918 (Part 2 of 2), John Murray, 1924

Hyde, Francis E., *Cunard and the North Atlantic, 1840–1973*, Macmillan Press, 1975

Jones, Captain John Treasure, (ed. Richard J. Tennant), *Tramp to Queen*, History Press, 2008.

Kaus, Gina, *Luxury Liner*, Cassell, 1932

Kerr, Michael, 'A Transatlantic Tale', *Daily Telegraph*, 4 February 2017

Layton, J. Kent, *The Edwardian Superliners: A Trio of Trios*, Amberley Publishing, 2012

Loos, Anita, *Gentlemen Prefer Blondes*, Penguin, 1998 (First pub. Boni and Liveright, 1925)

Lord, Walter, *A Night to Remember*, illus. edn Penguin, 1978 (First pub. Longmans, Green, 1956)

Lovell, Mary S., *The Riviera Set, 1920–1960*, Abacus, 2017

Mackrell, Judith, *The Unfinished Palazzo: Life, Love and Art in Venice*, Thames and Hudson, 2017

Margetson, Stella, *The Long Party*, Gordon Cremonesi, 1974

Mathair A'Chinn Suidhe – Trump's Mother, broadcast on BBC Alba, 17 September 2019 (in Gaelic, with English subtitles)

Maxtone-Graham, John, *The Only Way to Cross*, Patrick Stephens, 1983 (First pub. as *The North Atlantic Run*, Cassell, 1972)

Maxtone-Graham, John, *Crossing and Cruising: From the Golden Era of Ocean Liners to the Luxury Cruise Ships of Today*, Charles Scribner, 1992

Maxtone-Graham, John, *Normandie: France's Legendary Art Deco Ocean Liner*, W.W. Norton, 2007

Maxtone-Graham, John, *Titanic Tragedy: A New Look at the Lost Liner*, W.W. Norton, 2012

Maxtone-Graham, John (ed.), *Titanic Survivor: The Memoirs of Violet Jessop, Stewardess*, History Press, 2012

Maxwell, Elsa, *RSVP: Elsa Maxwell's Own Story*, Little, Brown, 1954

McAllister, Ian Hugh, *Lost Olympics: The Hilda James Story*, emp3books, 2013. See also www.lostolympics.co.uk

McCutcheon, Janette, *Cunard: A Photographic History*, Tempus Publishing, 2004

Miller, William H., *The Great Luxury Liners, 1927–1954*, Dover Publications, 1981

Miller, William H. and Hawley, Brian, *RMS Caronia: Cunard's Green Goddess*, History Press, 2011

Morton, Andrew, *Wallis in Love*, Michael O'Mara Books, 2018

Murphy, N.T.P., *In Search of Blandings*, Penguin, 1987

Murray, Roddy, 'Into the Silent Funeral: the Iolaire Disaster', *Stornoway Gazette*, 1994

Nixson, Maida M., *Ring Twice for the Stewardess*, John Long, 1954

Parris, Matthew (presenter), *Great Lives on Josephine Baker, with Mica Paris*, broadcast on BBC Radio 4, 18 May 2018

Piouffre, Gérard, *First Class: Legendary Ocean Liner Voyages Around the World*, Vendome Press, 2009

Powell, Violet, *The Life of a Provincial Lady: A Study of E.M. Delafield and her Works*, Heinemann, 1988

Preston, Diana, *Wilful Murder: The Sinking of the Lusitania*, Doubleday, 2002

Radic, Thérèse, *Melba: The Voice of Australia*, Palgrave Macmillan, 1986

Rhodes, Richard, *Hedy's Folly: The Life and Breakthrough Inventions of Hedy Lamarr*, Vintage Books 2011

Scott, Jeremy, *Women Who Dared To Break All the Rules*, Oneworld Publications, 2019

Sowerbutts, Edith, *Memoirs of a British Seaman* Private Papers, unpublished manuscript, Imperial War Museum

Spedding, Charles T., *Reminiscences of Transatlantic Travellers*, T. Fisher Unwin, 1926

Stanley, Jo, *From Cabin 'Boys' to Captains: 250 Years of Women at Sea*, History Press, 2016

Steel Ships and Iron Men (one of the series Reel History of Britain, episode 12 of 20), broadcast on BBC2, 20 September 2011

Stuart, Andrea, *Showgirls*, Jonathan Cape, 1996

Sykes, Christopher, *Nancy: The Life of Nancy Astor*, Panther Books, 1979

Taylor, D. J., *Bright Young People: The Lost Generation of London's Jazz Age*, Farrar, Straus and Giroux, 2007

Thorpe, D.R. (ed.), *Who's In, Who's Out: The Journals of Kenneth Rose, Volume One, 1944–1979*, Weidenfeld & Nicolson, 2018

Trayner, J.J. and Plumb, E.C., *Ship's Steward's Handbook*, Cunard Archives, shelving, Liverpool, SPEC 2015.a.001. Accessed 19 February 2018 (Small paperback book, listing duties and terminology for training ships stewards. Undated, probably early 1950s as it refers to government Acts etc. in the late 1940s)

Trumpington, Jean, *Coming Up Trumps: A Memoir*, Pan Books, 2015

Vanderbilt, Gloria and Furness, Lady Thelma, *Double Exposure: A Twin Autobiography*, Frederick Muller, 1959

Walker, Stanley, *Mrs Astor's Horse*, Frederick A. Stokes, 1935

Warwick, Sam and Roussel, Mike, *Shipwrecks of the Cunard Line*, History Press, 2012

Williams, A. Susan, *Ladies of Influence: Women of the Elite in Interwar Britain*, Penguin Books, 2001

Williams, David L. and de Kerbrech, Richard P., *Cabin Class Rivals*, History Press, 2015

Wills, Elspeth, *Stars Aboard*, Open Agency, 2003

Wilson, Christopher, *Dancing with the Devil: The Windsors and Jimmy Donahue*, HarperCollins, 2000

Woodcock, Roger, *Tyneside Cunarders*, Newcastle City Libraries and Arts, 1990

Woon, Basil, *The Frantic Atlantic: An Intimate Guide to the Well-Known Deep*, Alfred A. Knopf, 1927

Wright, Carol, *Cunard Cook Book*, J.M. Dent, 1969

Ziegler, Philip, *Diana Cooper*, Penguin Books, 1983

Ziegler, Philip, *Between the Wars: 1919–1939*, MacLehose Press, 2016

Index

About the Author

Cultural historian Siân Evans has worked for the National Trust, the V&A and the Design Museum, and is the author of several works of social history including *Queen Bees: Six Brilliant and Extraordinary Society Hostesses Between the Wars*, *Mrs Ronnie*, *The Manor Reborn* and *Life Below Stairs*. She lives in London.